ANATOMY RECALL
2nd Edition

RECALL SERIES EDITOR

LORNE H. BLACKBOURNE, M. D.
General Surgeon, Trauma and Critical Care Surgery
San Antonio, TX

ANATOMY RECALL
2nd Edition

Senior Editor

Jared L. Antevil, M.D.
General Surgeon
San Diego, California

Editor; Recall Series Editor

Lorne H. Blackbourne, M.D.
General Surgeon, Trauma and Critical Care Surgeon
San Antonio, Texas

CO-EDITOR, 1st Edition

Christopher Moore, M.D.
Emergency Medicine Physician
New Haven, Connecticut

LIPPINCOTT WILLIAMS & WILKINS
A **Wolters Kluwer** Company
Philadelphia • Baltimore • New York • London
Buenos Aires • Hong Kong • Sydney • Tokyo

Acquisitions Editor: Donna Balado
Managing Editor: Cheryl W. Stringfellow
Marketing Manager: Emilie Linkins
Production Editor: Sirkka E. H. Bertling
Designer: Holly McLaughlin
Compositor: Nesbitt Graphics, Inc.
Printer: R. R. Donnelley & Sons—Crawfordsville

**QS
18.2
A53653
2006**

351 West Camden Street
Baltimore, MD 21201

530 Walnut St.
Philadelphia, PA 19106

The publisher is not responsible (as a matter of product liability, negligence, or otherwise) for any injury resulting from any material contained herein. This publication contains information relating to general principles of medical care that should not be construed as specific instructions for individual patients. Manufacturers' product information and package inserts should be reviewed for current information, including contraindications, dosages, and precautions.

Printed in the United States of America

First Edition, 2000

Library of Congress Cataloging-in-Publication Data

Anatomy recall / senior editor, Jared L. Antevil.-- 2nd ed.
    p. ; cm. -- (Recall series)
  Includes index.
  ISBN 0-7817-9885-X
  1. Human anatomy--Examinations, questions, etc. I. Antevil, Jared, 1972- II. Title.
  III. Series.
  [DNLM: 1. Anatomy--Examination Questions. QS 18.2 A53653 2006]
  QM32.A656 2006
  611'.0076--dc22   2005016037

The publishers have made every effort to trace the copyright holders for borrowed material. If they have inadvertently overlooked any, they will be pleased to make the necessary arrangements at the first opportunity.

To purchase additional copies of this book, call our customer service department at **(800) 638-3030** or fax orders to **(301) 824-7390**. International customers should call **(301) 714-2324**.

Visit Lippincott Williams & Wilkins on the Internet: http://www.LWW.com.
Lippincott Williams & Wilkins customer service representatives are available from 8:30 am to 6:00 pm, EST.

05 06 07 08 09
1 2 3 4 5 6 7 8 9 10

Dedication

This book is dedicated to the medical students at the University of Virginia.

Preface

Anatomy Recall was written by medical students, physicians, and anatomists for use during a first-year gross anatomy course, as a review for the United States Medical Licensing Examination (USMLE) Step I, and for use during surgical clinical clerkships. The second edition has been expanded to include a wealth of clinical correlations, in order to emphasize relevant anatomic principles. The text has been supplemented with a basic overview of human embryology for use during preclinical coursework and for USMLE review. The added Surgical Anatomy Pearls section provides a forum for review immediately prior to time in the operating room, with high-yield surgical anatomy questions organized by specialty and operation.

While there are certainly many gross anatomy texts available, most are better suited for reference than for mastery of the basic anatomy required to be a successful medical student and physician.

Anatomy Recall is arranged in the extremely successful question-and-answer format that defines the entire *Recall* series—a format that emphasizes active acquisition of knowledge, rather than passive absorption of it. Where appropriate, simple figures have been included to supplement the text material. Each chapter concludes with a "Power Review"—indicated by this icon —that covers the most important and frequently tested facts in each subject area. These power reviews are ideal for a quick review prior to an anatomy examination, a board examination, or a surgery clerkship. Clinical pearls are accompanied by a stethoscope icon to facilitate a rapid review of clinically important anatomy during clerkships.

Anatomy is an exciting yet demanding course. It is important to have a text that is comprehensive yet readable and emphasizes (and reemphasizes) key points. A thorough initial study of anatomy will continue to reward you throughout a lifetime of clinical practice. It is our hope that *Anatomy Recall* will prove to be an invaluable tool for mastering the subject of anatomy. Good luck!

**ASSOCIATE EDITORS,
1st Edition, cont.
Clinton Nichols, M.D.**
Radiologist
San Diego, California

Ravi Rao, M.D.
Consultant
Cleveland, Ohio

Jeffrey Rentz, M.D.
General Surgeon, Resident in
    Cardiothoracic Surgery
Boston, Massachusetts

Peter Robinson, M.D.
Internist, Fellow in Cardiology
Denver, Colorado

John Schreiber, M.D.
Radiologist
McLean, Virginia

Albert Weed, M.D.
General Surgeon
Newport News, Virginia

**CONTRIBUTORS, 1st Edition
Wang Cheung, M.D.
Jamal Hairston, M.D.
Meredith LeMasters, M.D.
Steven Liu, M.D.
Bruce Lo, M.D.
Anu Meura, M.D.
Suzanne Perks, M.D.
Andrew Wang, M.D.
Thomas Wang, M.D.
Philip Zapata, M.D.**

Acknowledgments

The editors would like to acknowledge Emilie Linkins and Cheryl Stringfellow at Lippincott Williams & Wilkins for their help and vision in creating this revised and improved edition of *Anatomy Recall*.

Contents

1   Overview

It is important to adhere to a certain formalism when describing the location or movement of 1 body part relative to another; therefore, a significant portion of the anatomy course (like many introductory courses in medicine) is devoted to teaching a language necessary for communicating with other healthcare professionals.

ANATOMIC POSITION

What standard position is assumed when describing the human body?

That of a human standing facing forward, feet pointing forward, and palms facing anteriorly (the "anatomic position")

ANATOMIC PLANES

Describe the 3 basic anatomic planes.

1. **Horizontal (transverse):** A horizontal plane across the body in anatomic position; the most common cut used in computed tomography (CT) and magnetic resonance imaging (MRI)
2. **Sagittal:** A plane formed by a vertical midline cut that divides the body into right and left sides
3. **Coronal (frontal):** A plane formed by a cut across the body in anatomic position from side to side and top to bottom

Describe the median anatomic plane.

The imaginary line in the sagittal plane that bisects a structure into equal right and left halves

**Label the anatomic planes
on the figure below:**

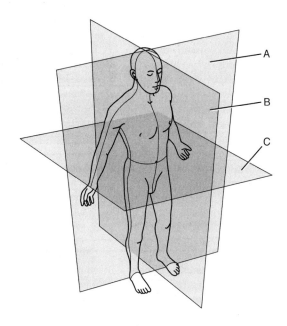

A = Coronal plane
B = Sagittal plane
C = Transverse plane

ANATOMIC DESCRIPTORS

Define the following terms:

Ventral/Anterior	Toward the front of the body
Dorsal/Posterior	Toward the back of the body
Medial	Closer to the midline
Lateral	Further from the midline
Superior	Closer to head
Inferior	Closer to feet

In the anatomic position, are the thumbs medial or lateral to the forefinger?	With the palms facing anteriorly, the thumbs are lateral to the other fingers.
Which is more superior, the umbilicus or the cervical spine?	The cervical spine
Define the following terms:	
Proximal	Closer to the center of the body (often considered the heart)
Distal	Further from the center of the body
Where is the wrist in relation to the shoulder?	The wrist is *distal* to the shoulder.
Which is more proximal, the femur or tibia?	The femur

ANATOMIC MOVEMENTS

Describe the innervation and characteristics of skeletal muscle.	Skeletal muscle is generally innervated by somatic nerves (i.e., movement is voluntary) and is located between 2 stable points (i.e., bones). Contraction results in movement of a structure.
What 4 parameters are used to describe skeletal muscles?	**Origin:** Usually the more proximal, more medial, and more stable structure that the muscle is attached to **Insertion:** Usually the more distal, more lateral structure that the muscle is attached to, and the 1 that is moved by contraction **Innervation:** The nerve that causes the muscle to contract **Action:** The result of the muscle contracting
Define the following muscle actions:	
Flexion	Decreasing the angle of a joint, or bending the joint

Extension	Increasing the angle of the joint, or straightening the joint
Abduction	Moving 1 structure away from another laterally (i.e., away from anatomic position)
Adduction	Moving 1 structure toward another medially (i.e., toward the anatomic position)—think add = together

What is meant by the anatomic terms "deep" and "superficial"?

Deep structures are further from the surface relative to superficial ones.

Describe the action that occurs with each of the following movements:

Kicking a soccer ball

Extension of the leg at the knee

Spreading the fingers

Abduction of the fingers at the metacarpophalangeal joints

Bringing an arm that is extended straight out and to the side laterally, toward the body

Adduction of the arm at the shoulder

What is the difference between ligaments and tendons?

Tendons attach the muscle to the bone, while ligaments attach bone to bone.

 What is a strain?

A partial or incomplete tear of a muscle or ligament

 What is a sprain?

A partial or incomplete tear of a tendon

2 The Head

THE SKULL

What is the skull?	The skeleton of the head, including the mandible
What are the 2 regions of the skull?	The neurocranium (i.e., the portion of the skull that encloses the brain) and the facial cranium
What is the calvaria?	The skull cap (i.e., the vault of the neurocranium, or the portion of the skull that is left when the facial bones are removed)
Which 4 bones contribute to the calvaria?	Frontal, parietal (2), occipital
What is diploë?	The spongy bone layer between the dense outer and inner bone layers of the calvaria
What are the 4 parts of the temporal bone?	Squamous (flat part articulating anteriorly with greater wing of sphenoid), petrous (dense, "stonelike," contributes to skull base), tympanic, and mastoid
What is the significance of the superior and inferior temporal lines?	They are attachment points for the temporalis muscle.
What region lies inferior to the temporal lines?	The temporal fossa
What is the name of the external opening in the temporal bone?	External auditory (acoustic) meatus
What bony process lies just posterior to the external auditory meatus?	Mastoid process

**Identify the structures on
the following lateral view of
the skull:**

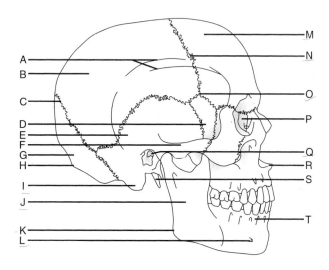

A = Inferior and superior temporal lines
B = Parietal bone
C = Lambdoid suture
D = Sphenoid bone, greater wing
E = Temporal bone
F = Zygomatic arch
G = Occipital bone
H = External occipital protuberance
I = Mastoid process
J = Ramus of the mandible
K = Angle of the mandible
L = Mental foramen
M = Frontal bone
N = Coronal suture
O = Pterion (the "p" is silent)
P = Lacrimal bone
Q = External auditory (acoustic) meatus
R = Anterior nasal spine
S = Styloid process
T = Alveolar process

**Identify the structures on
the following anterior view
of the skull:**

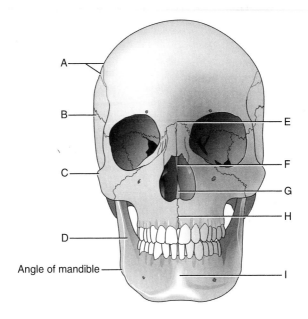

A = Temporal lines
B = Temporal fossa
C = Zygomatic arch
D = Ramus of mandible
E = Glabella
F = Perpendicular plate of ethmoid bone
G = Vomer (part of nasal septum)
H = Anterior nasal spine
I = Mandibular symphysis (line of union of mandibular halves)

**What is the clinical
significance of the communication between the middle
ear and the mastoid process?**

Severe middle ear infections (otitis media)
may spread to the mastoid process of the
temporal bone.

**What slender bony process
extends from the lower
portion of the petrous
temporal bone (in front of
the mastoid?)**

Styloid process

NEUROCRANIUM

Bones and Sutures

Which 8 bones make up the neurocranium?
The frontal bone, the 2 parietal bones, the 2 temporal bones, the occipital bone, the sphenoid bone, and the ethmoid bone

What are the immobile junctions between the bones of the neurocranium called?
Sutures

Which bones articulate at the:
    Coronal suture?
The frontal and parietal bones

    Sagittal suture?
The parietal bones of either side

    Lambdoid suture?
The parietal and occipital bones

What is the intersection of the lambdoid and sagittal sutures called?
The lambda

What is the intersection of the sagittal and coronal sutures called?
The bregma

What is a metopic suture?
A persistent frontal suture, present in approximately 2% of the population

What is craniosynostosis?
Premature closure of the sutures

What are fontanelles?
Large fibrous areas where several sutures meet; often called "soft spots" on an infant's head

What are the 2 largest fontanelles, and where are they located?
The anterior and posterior fontanelles, on the superior surface of the neurocranium

Which sutures form the borders of the posterior fontanelle?
The sagittal and lambdoid sutures

Which sutures form the borders of the anterior fontanelle?
The sagittal and coronal sutures

Identify the labeled points on the neurocranium on the following posterior and superior views:

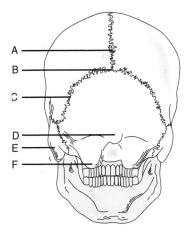

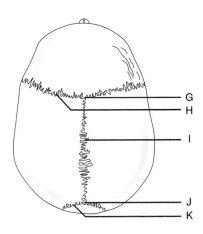

A = Sagittal suture
B = Lambda
C = Lambdoid suture
D = External occipital protuberance
E = Mastoid process
F = Occipital condyle
G = Bregma
H = Coronal suture
I = Sagittal suture
J = Lambda
K = Lambdoid suture

How can the anterior and posterior fontanelles be identified on an infant?

The anterior fontanelle is diamond-shaped and palpable in children younger than approximately 18 months. The posterior fontanelle is triangular and is not palpable past 1 year of age.

In adults, what is the name of the remnant of the:
Anterior fontanelle?

The bregma

Posterior fontanelle?

The lambda

What is the location of the anterolateral (sphenoidal) fontanelle called in adults?

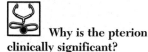 The pterion (Brain surgery using an anterolateral incision is called a "pterional approach.")

Which 4 bones articulate at the pterion?

Frontal, temporal, parietal, and sphenoid (greater wing). Note that the pterion is not a "spot," but rather an area about the size of a dime.

Why is the pterion clinically significant?

The thinnest part of the lateral skull, the pterion is vulnerable to fractures that can damage the middle meningeal artery, which lies on the internal skull surface in this region.

What is the name of the superior-most portion of the skull?

The vertex

INTERNAL SURFACE FEATURES

Label the following view of the floor of the neurocranium:

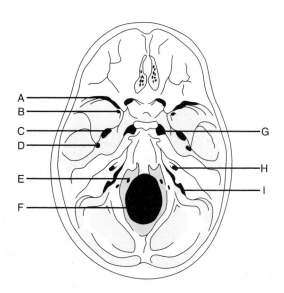

A = Superior orbital fissure
B = Foramen rotundum
C = Foramen ovale
D = Foramen spinosum
E = Hypoglossal canal
F = Foramen magnum
G = Foramen lacerum
H = Internal auditory meatus
I = Jugular foramen

Anterior Cranial Fossa

In addition to the ethmoid bone, which bones contribute to the floor of the anterior fossa?

The frontal bone and lesser wing of the sphenoid

What is the name of the transverse part of the ethmoid bone that lies anteriorly in the midline?

The cribriform plate. Fractures of this thin section of bone may lead to leakage of cerebrospinal fluid (CSF) from the nose ("CSF rhinorrhea").

What structure passes through the cribriform plate?

Cranial nerve (CN) I (the olfactory nerve)

What is the name of the sharp upward projection of the ethmoid bone in the midline?

The crista galli

What is the anatomic function of the crista galli?

It provides the anterior attachment site for the falx cerebri (i.e., the dural fold that lies in the longitudinal fissure between the 2 cerebral hemispheres).

What small midline foramen lies anterior to the crista galli?

Foramen cecum (leads into nasal cavity)

What 2 sets of small paired openings lie adjacent to the cribriform plate?

Anterior and posterior ethmoidal foramina

Middle Cranial Fossa

Which part of the brain occupies the middle cranial fossa?	The temporal lobes of the cerebral hemispheres
What are the borders of the middle cranial fossa:	
Anteriorly?	The greater and lesser wings of the sphenoid bones
Posteriorly?	The petrous part of the temporal bone
Laterally?	The squamous part of the temporal bone, the greater wings of the sphenoid bones, and the parietal bones
Ventrally (i.e., the floor)?	The temporal bones and the greater wings of the sphenoid bones
Which 3 structures pass from the middle cranial fossa into the orbit via the optic canal?	1. CN II (the optic nerve) 2. The ophthalmic artery (a branch of the internal carotid artery) 3. The central vein of the retina
Which opening between the greater and lesser wings of the sphenoid bone connects the middle cranial fossa with the orbit?	The superior orbital fissure
Which 6 structures pass from the middle cranial fossa to the orbit through the superior orbital fissure?	1. CN III (the oculomotor nerve) 2. CN IV (the trochlear nerve) 3. CN V_1 (the ophthalmic division of the trigeminal nerve) 4. CN VI (the abducens nerve) 5. The superior ophthalmic vein 6. The inferior ophthalmic vein
The foramen rotundum transmits structures between which 2 spaces?	The middle cranial fossa and the pterygopalatine fossa
Which structure passes through the foramen rotundum?	CN V_2 (the maxillary division of the trigeminal nerve)

The foramen ovale transmits structures between which 2 spaces?

The middle cranial fossa and the infratemporal fossa

Which 2 structures pass through the foramen ovale?

CN V_3 (the mandibular division of the trigeminal nerve) and the accessory meningeal artery

The foramen spinosum connects the middle cranial fossa with which space?

The infratemporal fossa (like the foramen ovale)

Which structure passes through the foramen spinosum?

The middle meningeal artery. Injury to this vessel due to skull fracture may lead to an "epidural hematoma."

What is the name of the opening that lies directly medial to foramen rotundum and foramen spinosum?

Foramen lacerum

The foramen lacerum lies at the junction of which cranial bones?

The sphenoid bone, the occipital bone, and the petrous part of the temporal bone

What structure is transmitted via the foramen lacerum?

Foramen lacerum is in fact not a true "foramen" at all. It is covered by a thin layer of cartilage, and does not actually transmit any structures.

Trace the route of exit for each division of the trigeminal nerve from the middle cranial fossa:

    CN V_1 (ophthalmic division)

To orbit via superior orbital fissure

    CN V_2 (maxillary division)

To pterygopalatine fossa via foramen rotundum

    CN V_3 (mandibular division)

To infratemporal fossa via foramen ovale

Grooves on the anterior part of the petrous temporal bone transmit which structures?

The greater and lesser petrosal nerves

What is the name of the thin plate of bone located at the junction of the petrous and squamous parts of the temporal bone?	The tegmen tympani
What is the clinical significance of this thin bone?	This bone, which separates the tympanic cavity from the middle cranial fossa, is so thin that infections of the middle ear can spread to the meninges and brain.
What is the name of the elevation of the sphenoid bone between the 2 optic canals?	The tuberculum sellae
What is the name of the depression posterior to the tuberculum sellae?	The sella turcica ("Turkish saddle")
What is the name of the bony ridge that defines the posterior limit of the sella turcica?	The dorsum sellae
What are the boundaries of the sella turcica: **Anteriorly?**	The tuberculum sellae
Posteriorly?	The dorsum sellae
Which organ lies in the hypophyseal fossa of the sella turcica?	The pituitary gland. Because of its anatomic location, pituitary tumors may impinge on the optic nerve (CN II), leading to visual field deficits.
Which space is located directly inferior to the sella turcica?	The sphenoid sinus (Surgery on the pituitary gland uses a "transsphenoidal" approach.)
Which structure forms the roof of the sella turcica?	The diaphragma sellae (i.e., 1 of the dural folds)

Which processes project from the lateral aspects of the dorsum sellae?

The posterior clinoid processes

What structure attaches to the posterior clinoid processes?

The tentorium cerebelli (i.e., the dural fold between the occipital lobes and the cerebellum)

Posterior Cranial Fossa

Which part of the brain lies in the posterior cranial fossa?

The cerebellum and brainstem

What are the borders of the posterior cranial fossa:
  Anteriorly?

The petrous part of the temporal bone and the basilar part of the occipital bone

  Posteriorly?

The occipital bone

  Ventrally (i.e., the floor)?

The occipital bone and the mastoid processes of the temporal bones

  Dorsally (i.e., the roof)?

The tentorium cerebelli (a dural fold that separates the cerebral hemispheres from the underlying cerebellum and brainstem)

Which 3 structures pass through the internal auditory meatus?

1. CN VII (the abducens nerve)
2. CN VIII (the vestibulocochlear nerve)
3. The labyrinthine artery

Which cranial foramen lies at the junction of the petrous part of the temporal bone and the occipital bone?

The jugular foramen

Which 6 structures pass through the jugular foramen?

1. CN IX (the glossopharyngeal nerve)
2. CN X (the vagus nerve)
3. CN XI (the accessory nerve)
4. The internal jugular vein (superior bulb)
5. The sigmoid sinus
6. The inferior petrosal sinus

Where is the hypoglossal canal in relation to the jugular foramen?

The hypoglossal canal lies just medial to the jugular foramen.

Which nerve passes through the hypoglossal canal?

CN XII (the hypoglossal nerve)

Which large opening lies in the posterior midline floor of the posterior fossa?

The foramen magnum. Edema of the brain after injury may lead to herniation of the brainstem and cerebellum from the posterior fossa into the spinal canal via this foramen ("uncal herniation").

Which structures pass through the foramen magnum?

1. The medulla oblongata (i.e., the lower aspect of the brainstem)
2. CN XI (the spinal accessory nerve)
3. The vertebral arteries
4. The venous plexus of the vertebral canal
5. The anterior and posterior spinal arteries

What is the name of the bony "ramp" just anterior to the foramen magnum?

The clivus

Which small opening may be present posterior to the foramen magnum?

The condyloid foramen

Which structure passes through the condyloid foramen?

The condyloid emissary vein

Which structures pass through the mastoid foramen?

The mastoid emissary vein and a branch of the occipital artery

What is the name of the midline crest on the inside of the occipital bone?

The internal occipital crest

Which structure attaches to this crest?

The falx cerebelli (i.e., the dural fold that separates the cerebellar hemispheres)

**Label the following features
of the external skull base:**

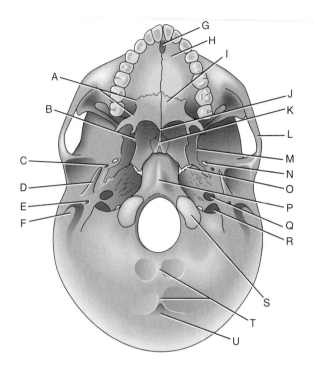

A = Greater and lesser palatine foramina
B = Medial pterygoid plate
C = Foramen spinosum
D = Styloid process
E = Stylomastoid foramen
F = Mastoid process
G = Incisive fossa
H = Palatine process of maxilla
I = Horizontal plate of palatine bone
J = Posterior nasal spine
K = Vomer
L = Zygomatic arch
M = Lateral pterygoid plate
N = Foramen ovale
O = Foramen lacerum
P = Pharyngeal tubercle

Q = Carotid canal
R = Jugular foramen
S = Occipital condyle
T = External occipital crest
U = External occipital protuberance

What is the superior termination of the internal occipital crest?

The internal occipital protuberance

Which structures are transmitted in the grooves that project laterally from the internal occipital protuberance along the occipital bone?

The transverse venous sinuses

What are the names of the external bony prominences located on either side of foramen magnum?

Occipital condyles

What is the function of the occipital condyles?

The occipital condyles articulate with the atlas (1st cervical vertebra), thereby supporting the skull on the spine.

What foramen lies just lateral to the occipital condyles?

Jugular foramen (transmits CN IX, X, XI, internal jugular vein, sigmoid sinus, and inferior petrosal sinus)

What foramen lies just posterior to the styloid process?

Stylomastoid foramen

What structure is transmitted by the stylomastoid foramen?

CN VII (facial nerve, exits skull)

FACIAL CRANIUM

What is the smooth median prominence of the frontal bone called?

The glabella

What bones form the prominence of the cheeks?

The zygomatic bones

Label the structures shown
on the following anterior
view of the skull:

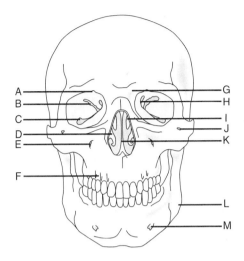

A = Supraorbital notch
B = Superior orbital fissure
C = Inferior orbital fissure
D = Inferior nasal concha
E = Infraorbital foramen
F = Alveolar process
G = Glabella
H = Optic canal
I = Middle nasal concha
J = Zygomaticofacial foramen
K = Nasal septum
L = Angle of the mandible
M = Mental foramen

Which bone forms the upper jaw?

The maxilla

What part of the mandible forms the prominence of the chin?

The mental protuberance

Which bone located between the orbits contains the cribriform plate and a perpendicular plate?

The ethmoid bone

Which structures pass through the:

Infraorbital foramen?

The infraorbital nerve (a continuation of CN V_2), the infraorbital artery, and the infraorbital vein

Zygomaticofacial foramen?

The zygomaticofacial nerve (branch of CN V_2) and zygomaticofacial vessels

Mental foramen?

The mental nerve (branch of CN V_3), the mental artery (termination of inferior alveolar branch of maxillary artery), and the mental vein

Orbit

Which bones form the margins of the orbit:

Superiorly?

The frontal bone (orbital plate)

Laterally?

The zygomatic bone, zygomatic process of the frontal bone, and greater wing of the sphenoid bone

Inferiorly?

The maxilla, palatine, and zygomatic bones. Fractures in this region may lead to an orbital "blow-out," where the contents of the orbit are displaced inferiorly.

Medially?

The ethmoid, lacrimal, sphenoid, and frontal bones

Which 2 fissures form a communication between the intracranial space and the orbit?

The superior orbital fissure (communicates with the middle cranial fossa) and the inferior orbital fissure (communicates with the infratemporal fossa)

Which structures pass into the orbit from the infratemporal fossa via the inferior orbital fissure?

The zygomatic branch of CN V_2 (the maxillary division of the trigeminal nerve) and the infraorbital artery

**Label the following diagram
of the bones of the orbit:**

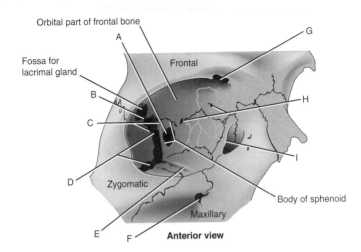

A = Optic canal (foramen)
B = "Lacrimal foramen"
C = Lesser and greater wings of
    sphenoid bone
D = Superior and inferior orbital fissures
E = Infraorbital groove
F = Infraorbital foramen
G = Supraorbital notch
H = Anterior and posterior ethmoidal
    foramina
I = Crest of lacrimal bone (posterior
    lacrimal crest)

**Which structures pass into
the orbit from the middle
cranial fossa via the superior
orbital fissure?**

CN III, IV, branches of V₁, VI, and the
superior/inferior ophthalmic veins

**What opening lies at the
apex of the orbit?**

The optic canal (transmits CN II, the
optic nerve)

**Which structures pass
through the supraorbital
notch (or foramen)?**

The supraorbital nerve (branch of CN V₁)
and the supraorbital vessels

Paranasal (Air) Sinuses

What are paranasal sinuses?

The paranasal sinuses are air spaces within the bones of the skull that communicate with the nasal cavity. Don't confuse the paranasal sinuses with the venous (dural) sinuses, which convey venous blood in the cranium.

List the 4 skull bones that have paranasal sinuses.

1. Frontal bone
2. Maxilla
3. Ethmoid bone
4. Sphenoid bone

What is the function of these sinuses?

Their function is unknown, although they are thought to lighten the skull and aid in resonation of the voice.

Which sinus, because of its location, can often lead to the spread of infection into the orbit?

The ethmoid sinus

Which sinus is susceptible to the spread of infection from a tooth?

The maxillary sinus. Roots of the posterior maxillary teeth often project up into this sinus.

Which sinuses may be present at birth?

The maxillary and ethmoid sinuses

Nasal Cavity

Which 4 bones form the roof of the nasal cavity?

The nasal bone, the frontal bone, the cribriform plate of the ethmoid bone, and the body of the sphenoid bone

Which bones form the floor of the nasal cavity (and the hard palate, or the anterior portion of the roof of the mouth)?

The palatine process of the maxilla and the horizontal plate of the palatine bone

Which 8 bones contribute to the lateral wall of the nasal cavity?

1. The ethmoid bone
2. The medial pterygoid plate

3. The perpendicular plate of the palatine bone
4. The maxilla
5. The nasal bone
6. The frontal bone
7. The lacrimal bone
8. The inferior concha

Which opening forms a communication between the nasal cavity and the nasopharynx?

The choanae, a large opening at the back of the nasal cavity

Which structure divides the nasal cavity into left and right sides?

The nasal septum

Which 2 bones make up the rest of the nasal septum?

The vomer and the perpendicular plate of the ethmoid bone (also very small contributions from maxilla and palatine bone)

What are the 3 bony projections from the lateral nasal wall called?

The nasal conchae (or turbinates). The superior and middle conchae are part of the ethmoid bone, while the inferior nasal concha is an independent bone.

What are the spaces below each of the nasal conchae called?

Meatuses (e.g., the superior meatus is the space between the superior and middle conchae; the middle meatus is the space between the middle and inferior conchae; and the inferior meatus is the space inferior to the inferior concha)

The inferior meatus contains the opening to which structure?

The nasolacrimal duct

What is the name of the space posterosuperior to the superior concha?

The sphenoethmoid recess

What is the rounded prominence on the wall of the middle meatus?

The ethmoid bulla

What is the hiatus semilunaris?	The curved cleft inferior to the ethmoid bulla
What is the infundibulum?	The channel at the anterior aspect of the hiatus semilunaris
Describe the drainage of each of the following paranasal sinuses:	
The anterior ethmoid	The hiatus semilunaris (via the infundibulum), located in the middle nasal meatus
The middle ethmoid sinus	The ethmoid bulla, located in the middle nasal meatus
The posterior ethmoid sinus	The superior nasal meatus
The frontal sinus	The middle nasal meatus (via the frontonasal duct, which opens into the infundibulum)
The sphenoid sinus	The sphenoethmoidal recess
The maxillary sinus	The hiatus semilunaris
Which 3 arteries supply branches to the nasal cavity?	The maxillary and facial arteries (branches of the external carotid artery) and the ophthalmic artery (a branch of the internal carotid artery)
What are the 2 primary branches of the ophthalmic artery that supply the nasal cavity?	The anterior and posterior ethmoidal arteries, which supply the lateral wall and nasal septum
What are the 2 branches of the maxillary artery that supply the nasal cavity?	The sphenopalatine artery (which supplies the conchae, meatus, and nasal septum) and the descending palatine artery (which also supplies the nasal septum)
Which branch of the facial artery supplies the nasal cavity?	The superior labial artery

Mandible

Name the 2 processes on the superior aspect of the ramus.

The coronoid process (anterior) and the condyloid process (posterior)

What is the name of the notch located between the coronoid and condyloid processes?

The mandibular notch

What is the name of the opening on the medial surface of the mandibular ramus?

The mandibular foramen

What opening does the mandibular foramen lead to?

The mandibular canal

What structures lie in the mandibular canal?

The inferior alveolar nerves and vessels

Label the following figure of the mandible:

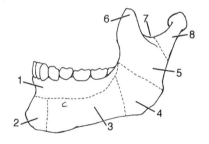

1 = Alveolar process
2 = Symphysis
3 = Body
4 = Angle
5 = Ramus
6 = Coronoid process
7 = Mandibular notch
8 = Condylar process

Infratemporal Fossa

Which 2 muscles lie within the infratemporal fossa?

The medial and lateral pterygoid muscles (both muscles of mastication)

What major artery lies within the infratemporal fossa?	The maxillary artery
Which nerves lie within the infratemporal fossa?	The chorda tympani nerve (a branch of CN VII within the temporal bone) and branches of CN V$_3$ (mandibular division of the trigeminal nerve)

Which structures pass through the:

Petrotympanic fissure?	The chorda tympani (a branch of the facial nerve within the temporal bone)
The stylomastoid foramen?	The facial nerve (CN VII)
The greater palatine foramen?	The greater palatine nerve and vessels
The lesser palatine foramen?	The lesser palatine nerve and vessels

THE SCALP AND SUPERFICIAL AND DEEP FACE

SCALP

What is the scalp?	The skin and fascia that covers the neurocranium
What are the 5 layers of the scalp?	**SCALP** **S**kin **C**onnective tissue **A**poneurosis (galea aponeurotica) **L**oose connective tissue **P**ericranium
Which scalp layer provides the scalp's strength?	The aponeurosis ("galea")
Which layer contains nerves and blood vessels?	The connective tissue layer
Branches of which artery constitute the major blood supply of the scalp?	The external carotid artery

Which branches of the external carotid artery supply the scalp?	The superficial temporal, posterior auricular, and occipital arteries
Which branches of the *internal* carotid artery supply the scalp?	The supratrochlear and supraorbital arteries (via the ophthalmic artery)
What feature(s) of scalp arteries can predispose them to profuse bleeding after injury?	The scalp vessels connect with each other via an extensive "anastomotic" network. Additionally, they are held in place by dense connective tissue, which may prevent them from retracting (a normal protective mechanism to reduce bleeding from an artery).
What is unique about the veins of the scalp?	They have no valves.
What connects the veins of the scalp with the veins of the skull bones and the veins within the cranium?	Emissary veins
What nerves provide sensory innervation to the scalp?	**From CN V$_1$:** Supraorbital and supratrochlear nerves **From CN V$_2$:** Zygomaticotemporal nerve **From CN V$_3$:** Auriculotemporal nerve **From C2 spinal nerve:** Greater occipital nerve and spinal nerves C2 and C3 **From C3 spinal nerve:** 3rd occipital nerve

NOSE

What is the medical term for nostrils?	Nares
What is the name of the cartilaginous part of the external nose that surrounds each naris?	The ala nasi ("wing of the nose")

What is the dilated part of the nostril called?

The vestibule

What 3 effects does the nose have on inspired air?

Warming, moistening, and filtering

What is the source of blood supply to the nose?

1. **Internal carotid artery:** Anterior and posterior ethmoidal arteries via ophthalmic artery
2. **External carotid artery:** Superficial labial artery (via facial artery) and sphenopalatine artery (via internal maxillary artery)

Name the 3 bones that make up the posterior nasal septum.

1. Ethmoid (perpendicular plate)
2. Vomer (Latin for "plow")
3. Palatine (Some also include maxillary crest)

MUSCLES OF FACIAL EXPRESSION

Label the muscles of facial expression on the following figure:

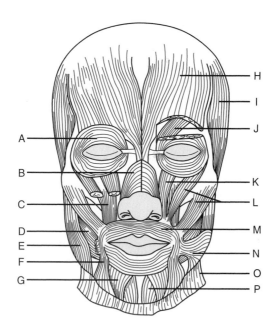

A = Orbicularis oculi
B = Nasalis
C = Levator anguli oris
D = Buccinator
E = Masseter
F = Depressor anguli oris
G = Depressor labii inferioris
H = Frontalis
I = Temporalis
J = Corrugator (supercilii)
K = Levator labii superioris
L = Zygomaticus major and minor
M = Orbicularis oris
N = Risorius
O = Platysma
P = Mentalis

Dysfunction of which muscle results in difficulty with blinking?

The orbicularis oculi (closes eyes)

Which muscle is primarily responsible for raising the eyebrows?

The frontalis muscle

Which muscle elevates the corners of the mouth?

Levator anguli oris

Which muscle depresses the corners of the mouth?

Depressor anguli oris

What is the action of the buccinator muscle?

It aids in chewing ("mastication") by pressing the cheeks against the molars.

The buccinator muscle is pierced by what structure?

The parotid duct en route to the oral cavity

Which muscle elevates the upper lip?

Levator labii superioris

What is the action of the mentalis muscle?

It raises the chin of the skin and protrudes the lower lip.

Which muscle encircles the mouth?

Orbicularis oris

What large, flat facial muscle extends into the neck?

The platysma muscle. Because none of the critical neck structures lies superficial to this muscle, penetrating injuries to the neck are only considered significant if they extend through the platysma.

Which nerve innervates the muscles of facial expression?

CN VII (the facial nerve)

TEMPOROMANDIBULAR JOINT (TMJ) AND MUSCLES OF MASTICATION

Temporomandibular Joint (TMJ)

What type of joint is the TMJ?

A synovial joint

Which 2 types of movements are provided by the TMJ?

Hinge movement and sliding movement

What are the articular surfaces of the TMJ?

The articular tubercle and mandibular fossa of the temporal bone and the condyloid process of the mandible

Name the 3 ligaments of the TMJ.

1. The lateral temporomandibular ligament
2. The sphenomandibular ligament
3. The stylomandibular ligament

Which ligament reinforces the TMJ by stretching from the tubercle on the zygoma to the neck of the mandible?

The lateral temporomandibular ligament

Which ligament reinforces the TMJ by stretching from the spine of the sphenoid bone to the lingula (midinner ramus) of the mandible?

The sphenomandibular ligament

Dislocation of the TMJ usually occurs in which direction?

Anteriorly

What is the "TMJ syndrome"?

A common disorder characterized by pain at the TMJ (usually unilateral) due to a variety of causes. Treatment is usually conservative, with heat packs, soft diet, and antiinflammatory medications.

Muscles of Mastication

What are the 4 muscles of mastication?

1. The masseter muscle
2. The temporalis muscle
3. The medial pterygoid muscle
4. The lateral pterygoid muscle

Describe the origin and insertion for each of the following:

Masseter muscle

Origin: The lower border and medial surface of the zygomatic arch
Insertion: The lateral surface of the coronoid process, ramus, and angle of the mandible

Temporalis muscle

Origin: The floor of the temporal fossa
Insertion: The coronoid process and ramus of the mandible

Lateral pterygoid muscle (superior head)

Origin: The infratemporal surface of the sphenoid bone
Insertion: The articular disk and capsule of the TMJ

Lateral pterygoid muscle (inferior head)

Origin: The lateral surface of the lateral pterygoid plate
Insertion: The neck of the mandible

Medial pterygoid muscle

Origin: The tuber of the maxilla, the medial surface of the lateral pterygoid plate, and the pyramidal process of the palatine bone
Insertion: The medial surface of the angle and the ramus of the mandible

Label the muscles of mastication:

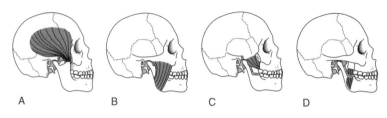

A   B   C   D

A = Temporalis
B = Masseter
C = Lateral pterygoid
D = Medial pterygoid

Which muscle is responsible for:

    Closing the jaw and protruding the mandible?

The medial pterygoid muscle

    Opening the jaw and protruding the mandible?

The lateral pterygoid muscle

Which muscle is the primary actor in:

    Depressing the mandible at the TMJ?

The lateral pterygoid muscle

    Retracting the mandible?

The temporalis muscle

Which 3 muscles act to elevate the mandible at the TMJ?

The temporalis, masseter, and medial pterygoid muscles

Which 2 muscles act to close the jaw and retract the mandible?

The masseter muscle and the temporalis muscle

Which arteries supply the muscles of mastication?

Small branches from the maxillary artery (sometimes called the pterygoid branches)

PAROTID GLAND

What is the name of the largest of the 3 sets of salivary glands?	The parotid gland
What are the boundaries of the space containing the parotid gland:	
Anteriorly?	The mandible and muscles of mastication
Posteriorly?	The external auditory meatus and mastoid process
Medially?	The styloid process
Superiorly?	The zygomatic arch

Which structure divides the parotid gland into superficial and deep parts?

CN VII (the facial nerve). This nerve serves as a surgical landmark and is the deep limit of resection in a "superficial parotidectomy" (performed for most benign parotid tumors).

Which blood vessels are located at the upper pole of the parotid, anterior to the ear?

The superficial temporal artery and vein

The superficial temporal vessels are accompanied by what nerve?

The auriculotemporal nerve. Injury to this nerve commonly occurs during parotid surgery, and may lead to cross-innervation with nearby sympathetic fibers. This can lead to Frey's syndrome, a clinical phenomenon where the sight/smell of food leads to flushing and perspiration (aka "crocodile tears").

What is the name of the major duct of the parotid gland?

The parotid duct (Stensen's duct)

Where does the parotid duct begin?	At the anterior aspect of the parotid gland
The parotid duct pierces which muscle on its course anteriorly?	The buccinator
Where does the parotid duct open into the oral cavity?	Opposite the second upper molar tooth
Which 3 major structures traverse the parotid gland?	1. CN VII (the facial nerve) 2. The retromandibular vein 3. The external carotid artery
From which ganglion do the parasympathetic fibers that supply the parotid gland originate?	The otic ganglion
What is the source of parasympathetic fibers to the otic ganglion?	Parasympathetic fibers originate in the inferior salivary nucleus of CN IX (the glossopharyngeal nerve), follow the tympanic branch, and then travel to the otic ganglion via the lesser petrosal nerve.
Which nerve transmits postganglionic parasympathetic fibers to the parotid gland?	After passing through the otic ganglion, the postganglionic parasympathetic fibers are transmitted to the parotid gland via the auriculotemporal nerve.
Sympathetic innervation to the parotid gland follows which structure?	The external carotid artery
What intraoperative technique is most frequently used to locate the facial nerve within the parotid gland?	Identification of the nerve as it exits the skull at the stylomastoid foramen. From this point, the nerve can be traced distally into the gland.

SUBMANDIBULAR REGION

Name the 2 sets of salivary glands that lie in the submandibular region.	1. Submandibular glands 2. Sublingual glands

Of the 3 sets of salivary glands, which is the smallest?	The sublingual glands
Where do the ducts of the sublingual salivary glands open into the mouth?	The mucous membrane of the floor of the mouth
Which structure separates the parotid gland and the submandibular glands?	The stylomandibular ligament
Which duct drains the submandibular gland?	The submandibular ("Wharton's") duct
The submandibular duct lies between which 2 structures?	The sublingual glands and the genioglossus muscle
Where does the submandibular duct empty?	Just lateral to the frenulum of the tongue
Which major blood vessels run in the submandibular region?	The facial and lingual arteries (both branches of the external carotid artery)
What is the source of parasympathetic innervation to the submandibular and sublingual salivary glands?	CN VII (the facial nerve), via the chorda tympani and lingual nerve

INNERVATION OF THE FACE

Which cranial nerve provides motor innervation to the face?	CN VII (the facial nerve)
How many terminal branches does CN VII have?	5
By what route does CN VII exit the skull?	Via the stylomastoid foramen
Within which structure does CN VII divide into its branches?	The parotid gland

Identify the 5 terminal branches of CN VII on the following figure:

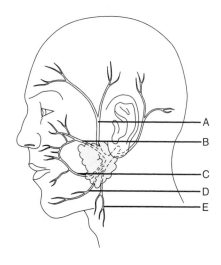

A = Temporal branch
B = Zygomatic branch
C = Buccal branch
D = Mandibular branch
E = Cervical branch

What clinical syndrome is characterized by facial nerve paralysis?

Idiopathic dysfunction of the facial nerve is termed Bell's palsy.

Which cranial nerve provides sensory innervation to the face?

CN V (the trigeminal nerve)

What are the 3 divisions of the trigeminal nerve?

Ophthalmic, maxillary, and mandibular

What 2 branches of the cervical plexus innervate the skin behind the ear?

Lesser occipital and great auricular (both from C2, C3)

The supraorbital and supratrochlear nerves are terminal branches of which nerve from the ophthalmic division of the trigeminal nerve?

The frontal nerve

Identify the nerves that provide sensory innervation to the face on the following figure:

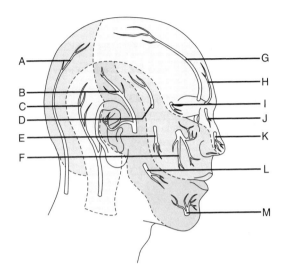

A = Greater occipital nerve
B = Auriculotemporal nerve (V₂)
C = Lesser occipital nerve
D = Zygomaticotemporal nerve (V₂)
E = Zygomaticofacial nerve (V₂)
F = Infraorbital nerve (V₂)
G = Supraorbital nerve (V₁)
H = Supratrochlear nerve (V₁)
I = Lacrimal nerve (V₁)
J = Infratrochlear nerve (V₁)
K = External nasal nerve (V₁)
L = Buccal nerve (V₃)
M = Mental nerve (V₃)

What areas are innervated by the supraorbital and supratrochlear nerves?

The scalp, forehead, and upper eyelid

Which nerve provides sensory innervation to the eye and the septum, lateral walls, and tip of the nose?

The nasociliary nerve, a branch of CN V₁ (the ophthalmic division of the trigeminal nerve)

Which branch of the nasociliary nerve provides sensory innervation to the septum, lateral walls, and tip of the nose?

The anterior ethmoidal nerve

What are the 3 major cutaneous branches of CN V_3 (the mandibular division of the trigeminal nerve)?

The auriculotemporal, buccal, and mental nerves

Which cranial nerve provides special sensory innervation to the nose?

CN I (the olfactory nerve). CN I passes through openings of the cribriform plate of the ethmoid bone on the way to the olfactory bulbs.

Describe the sensory innervation of the:
    Roof of the mouth

The greater palatine and nasopalatine nerves (branches of CN V_2, the maxillary division of the trigeminal nerve)

    Floor of the mouth

The lingual nerve (a branch of CN V_3, the mandibular division of the trigeminal nerve)

    Cheek

The buccal nerve (also a branch of CN V_3)

VASCULATURE OF THE FACE

Arteries

What is the source of arterial blood to the face?

The external carotid artery

Name the 8 branches of the external carotid artery, from proximal to distal.

1. Superior thyroid artery
2. Ascending pharyngeal artery
3. Lingual artery
4. Facial artery
5. Occipital artery
6. Posterior auricular artery
7. Superficial temporal artery
8. Maxillary artery

Facial Artery

Describe the course of the facial artery in the submandibular region.

The facial artery arises from the external carotid artery superior to the hyoid bone, ascends deep to the digastric and stylohyoid muscles and then deep to the submandibular gland, hooks around the inferior border of the mandibular body, and then enters the anterior margin of the masseter muscle.

What cranial nerve branch crosses over the facial artery at the lower border of the mandible?

The mandibular branch of CN VII (facial nerve)

Where can the pulse of the facial artery be easily palpated?

Just inferior to the mandible at the anterior border of the masseter muscle

Name 3 branches of the facial artery.

1. Inferior labial artery
2. Superior labial artery
3. Lateral nasal artery

The facial artery terminates as what vessel?

Angular artery

Maxillary Artery

Where does the maxillary artery branch from the external carotid artery?

At the posterior border of the ramus of the mandible, within the parotid gland

The maxillary artery is divided into 3 parts by which muscle?

The lateral pterygoid muscle

Where does the pterygopalatine part of the maxillary artery run in relation to the lateral pterygoid muscle?

Between the 2 heads

Identify the branches of the
maxillary artery on the
following figure:

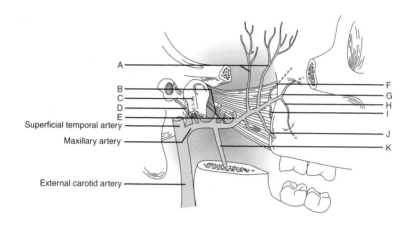

A = Deep temporal artery
B = Middle meningeal artery
C = Anterior tympanic artery
D = Deep auricular artery
E = Masseter artery
F = Sphenopalatine artery
G = Infraorbital artery
H = Posterior superior alveolar artery
I = Descending palatine artery
J = Buccal artery
K = Inferior alveolar artery

**Name the 5 branches of the
maxillary artery within the
infratemporal fossa.**

1. Deep auricular artery
2. Anterior tympanic artery
3. Middle meningeal artery
4. Accessory meningeal artery
5. Inferior alveolar artery

Which branch of the maxillary artery supplies the:

External auditory meatus?   The deep auricular artery

Tympanic membrane?   The anterior tympanic artery

Damage to which artery may result in an epidural hematoma?

The middle meningeal artery (most often damaged in fractures of the temporal bone)

How does the middle meningeal artery enter the skull?

Through the foramen spinosum

Which branch of the maxillary artery supplies the chin and lower teeth?

The inferior alveolar artery

Within what region does the maxillary artery terminate?

The pterygopalatine fossa

Name the 6 major branches of the maxillary artery in the pterygopalatine fossa.

1. Posterior superior alveolar artery
2. Infraorbital artery
3. Descending palatine artery
4. Artery of the pterygoid canal
5. Pharyngeal artery
6. Sphenopalatine artery

Describe the course of the sphenopalatine artery and the structures it supplies.

The sphenopalatine artery leaves the pterygopalatine fossa, passes through the sphenopalatine foramen, and enters the nasal cavity, where it supplies the conchae, meatus, and nasal septum.

Superficial Temporal Artery

Where can the superficial temporal arterial pulse be palpated?

Just anterior to the auricle of the external ear

Which nerve accompanies the superficial temporal artery?

The auriculotemporal nerve

The superficial temporal artery gives rise to the transverse facial artery. Between which 2 structures does the transverse facial artery pass?

The zygomatic arch superiorly and the parotid duct inferiorly

Veins

Describe 3 pathways for venous drainage in the face and scalp.	1. Facial vein to the retromandibular vein to the external jugular vein 2. Plexuses within the face to the external jugular vein 3. Venous (dural) sinuses to the internal jugular vein
Describe the origin of the retromandibular vein.	The retromandibular vein is formed when the superficial temporal and maxillary veins unite.
Describe the origin of the facial vein.	The supraorbital and supratrochlear veins join to form the angular vein, which becomes the facial vein at the lower margin of the orbit.
The facial vein joins with which structure to form the internal jugular vein?	The retromandibular vein

THE ORAL CAVITY

PALATE

What forms the roof of the mouth?	The hard palate (anteriorly) and the soft palate (posteriorly)
What is the name of the midline structure that hangs from the soft palate?	The uvula
Which bones comprise the hard palate?	The palatine processes of the maxilla and the horizontal plates of the palatine bones
What clinical condition may result from abnormal development of the lip and palate?	Clefts of the lip and palate are the most commonly encountered congenital malformations of the head and neck. They often result in severe functional deficits of speech and mastication. Treatment is surgical.
Where is the incisive foramen located?	Posterior to the central incisor teeth

Which structures pass through the incisive foramen?	The greater palatine artery (branch of the sphenopalatine artery) and the nasopalatine nerve

TEETH

What is the normal number of adult teeth?	32 (8 incisors, 4 canines, 8 premolars, and 12 molars)
What is the normal number of deciduous (baby) teeth?	20
Which nerves provide innervation to the maxillary teeth?	The anterior, middle, and posterior superior alveolar branches of CN V_2 (the maxillary division of the trigeminal nerve)
Which nerve innervates the mandibular teeth?	The inferior alveolar branch of CN V_3 (the mandibular nerve)

TONGUE

What is the name of the mucous membrane fold on the midline undersurface of the tongue?	The frenulum
What is ankyloglossia?	An abnormally short frenulum (can lead to speech impediment)
What duct enters the mouth on either side of the frenulum, beneath the tongue?	The submandibular salivary duct (aka Wharton's duct)
Where does the parotid duct (Stensen's) enter the mouth?	Adjacent to the 2nd upper molar
What nerve provides for taste sensation on the anterior two thirds of the tongue?	The chorda tympani, a branch of CN VII (the facial nerve)
What other structures are innervated by the chorda tympani?	The chorda tympani provides parasympathetic innervation to the submandibular and sublingual salivary glands and to the lacrimal glands.

Which nerve provides for taste sensation on the posterior third of the tongue?

CN IX (the glossopharyngeal nerve)

Which nerves provide tactile sensory innervation to the tongue?

Anterior two thirds—CN V_3 (the mandibular division of the trigeminal nerve) Posterior one-third—CN IX

Musculature of the Tongue

What is the function of the *intrinsic* tongue muscles?

They help the tongue maintain its shape.

Which 4 muscles comprise the *extrinsic* musculature of the tongue?

1. Genioglossus
2 . Hyoglossus
3. Styloglossus
4. Palatoglossus (Some do not consider this a tongue muscle, but rather associate it with the soft palate.)

What are the origin, insertion, and action of the:
   Genioglossus muscle?

Origin: The genial tubercle of the mandible
Insertion: The inferior aspect of the tongue and the body of the hyoid bone
Action: Protrudes and depresses the tongue

   Hyoglossus muscle?

Origin: The body and greater horn of the hyoid bone
Insertion: The side and inferior aspect of the tongue
Action: Depresses and retracts the tongue

   Styloglossus muscle?

Origin: The styloid process
Insertion: The side and inferior aspect of the tongue
Action: Retracts and elevates tongue

   Palatoglossus muscle?

Origin: Aponeuroses of the soft palate
Insertion: The dorsolateral side of the tongue
Action: Elevates the tongue

Which 1 of the extrinsic muscles of the tongue is not innervated by CN XII (the hypoglossal nerve)?

The palatoglossus (This muscle is innervated by CN XII [the vagus nerve] via the pharyngeal plexus.)

Lesions of CN XII cause the tongue to deviate toward which side?

Toward the side of the lesion (This is known as the "wheelbarrow effect"; think of what happens when you push a wheelbarrow with 1 hand—to which side does it tend to deviate?)

Which muscles act to:
    Retract the tongue?

Hyoglossus and styloglossus

    Elevate the tongue?

Styloglossus and palatoglossus

    Protrude the tongue?

Genioglossus

    Depress the tongue?

Genioglossus and hyoglossus

Vasculature of the Tongue

What is the arterial supply to the tongue?

The lingual branch of the external carotid artery, the ascending pharyngeal artery, and branches of the facial artery

What is the lymphatic drainage from the:
    Anterior third (tip) of the tongue?

The submental nodes

    Posterior two thirds of the tongue?

The submental nodes and the submandibular nodes (to the deep cervical lymphatic chain)

What are the lymph nodules located under the posterior tongue called?

The lingual tonsils

THE PHARYNX

What is the pharynx?

A muscular tube through which food and water pass to the esophagus and air passes to the larynx, trachea, and lungs

PHARYNGEAL MUSCLES

What 2 groups of muscles comprise the pharynx?

1. External circular layer (the constrictors)
2. Internal longitudinal layer

Pharyngeal Constrictor (External) Muscles

Name the 3 constrictor muscles, from interior to exterior.

1. Superior constrictor
2. Middle constrictor
3. Inferior constrictor

What is the action of the constrictors?

By constricting in a coordinated fashion, these muscles push food into the esophagus. Constriction is under autonomic (involuntary) control.

What is the origin of the:

Superior constrictor?

The pterygoid hamulus, pterygomandibular raphe, and posterior mylohyoid line of the mandible

Middle constrictor?

The stylohyoid ligament and hyoid bone

Inferior constrictor?

The thyroid and cricoid cartilages of the larynx

What is the common insertion for the 3 constrictor muscles?

The median raphe of the pharynx (i.e., the midline in the posterior of the pharynx)

What nerve innervates all of the constrictor muscles?

The pharyngeal and superior laryngeal branches of CN X (the vagus nerve), via the pharyngeal plexus. Difficulty swallowing (dysphagia) is a classic symptom of a surgical injury to the vagus nerve in the neck.

Longitudinal (Internal) Muscles

Name the 3 longitudinal (internal) muscles of the pharynx.

1. Palatopharyngeus muscle
2. Salpingopharyngeus muscle
3. Stylopharyngeus muscle

What is the action of the longitudinal muscles?

These muscles raise the pharynx and larynx during swallowing and speaking.

What is the origin of the:

Palatopharyngeus muscle?

The hard palate (palato-)

Salpingopharyngeus muscle?

The cartilaginous auditory (eustachian) tube (Salpinx means "tube" in Latin.)

Stylopharyngeus muscle?   The styloid process (stylo-)

What is the common insertion of the longitudinal muscles?   The posterior and superior border of the thyroid cartilage

What is the innervation of the longitudinal muscles?   The palatopharyngeus muscle and the salpingopharyngeus muscle receive innervation from CN X (the vagus nerve) via the pharyngeal plexus. The stylopharyngeus muscle is innervated by CN IX (the glossopharyngeal nerve).

PHARYNGEAL REGIONS

Identify the labeled structures on the following posterior view of the pharyngeal region:

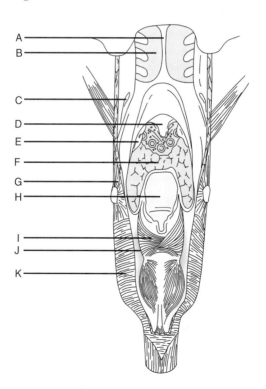

A = Nasal septum
B = Nasal cavity
C = Eustachian (auditory) tube
D = Uvula
E = Tonsil
F = Tongue
G = Middle constrictor muscle
H = Epiglottis
I = Arytenoid muscle
J = Piriform recess
K = Inferior constrictor muscle

What are the 3 divisions of the pharynx?

The nasopharynx, oropharynx, and laryngopharynx (hypopharynx)

Nasopharynx

What structure forms the boundary between the nasal cavity and the nasopharynx?

The choanae, a large opening at the posterior extent of the nasal cavity

What is Waldeyer's ring?

A ring of lymphoid tissue in the posterior oropharynx, consisting of the palatine, lingual, and pharyngeal tonsils.

Recurrent infections/inflammation of this tissue may necessitate removal (tonsillectomy).

Which structure connects the tympanic cavity with the nasopharynx?

The auditory (eustachian tube). This tube has a role in regulating pressure within the middle ear. Its obstruction or dysfunction may impair hearing. (You force this tube open when you "pop" your ears on an airplane.)

Where in the nasopharynx is the opening of the auditory (eustachian) tube?

In the lateral wall

Which muscle attaches to the auditory (eustachian) tube?

The salpingopharyngeus muscle

Oropharynx

What forms the border between the oral cavity and the oropharynx?

The palatoglossal arch (formed by underlying muscle of the same name)

What are the superior and inferior borders of the oropharynx?

The soft palate (superiorly) and the superior border of the epiglottis (inferiorly)

Which 2 folds bound the oropharynx laterally?

The palatoglossal and palatopharyngeal arches

What lies between these 2 folds?

The palatine tonsils (within the tonsillar sinus)

What is the name of the inferior projection from the midline of the soft palate?

The uvula (aka "that little thing at the back of your throat")

What is the sensory innervation of the oropharynx?

CN IX (the glossopharyngeal nerve)

List the 5 muscles of the soft palate.

1. Tensor veli palatini muscle
2. Levator veli palatini muscle
3. Palatoglossus muscle
4. Palatopharyngeus muscle
5. Musculus uvulae muscle

Which muscle is responsible for elevating and retracting the soft palate?

The levator veli palatini

Laryngopharynx (Hypopharynx)

What is the laryngopharynx?

The portion of the pharynx that lies posterior to the larynx

What are the superior and inferior margins of the laryngopharynx?

Superior: The upper border of the epiglottis
Inferior: The lower border of the cricoid cartilage (i.e., the beginning of the trachea)

THE EYE AND ADNEXA

EYEBALL

Identify the structures of the eye on the following figure:

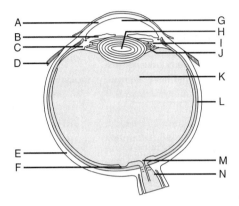

A = Cornea
B = Iris
C = Ciliary body
D = Lateral rectus muscle
E = Sclera
F = Macula (a part of the retina)
G = Aqueous humor
H = Lens
I = Ciliary processes
J = Suspensory ligament (zonular fibers)
K = Vitreous body
L = Choroid
M = Optic disc
N = CN II (the optic nerve)

CHAMBERS OF THE EYE

The lens lies between which 2 structures?

The iris (anteriorly) and the vitreous body (posteriorly). The lens functions to transmit light. Loss of its transparency is common with age (cataracts) and may necessitate lens replacement to preserve vision.

The iris divides the space between the lens and cornea into which chambers?

The anterior and posterior chambers

What fills these chambers?

Aqueous humor

Trace the flow of aqueous humor.

Made by the ciliary processes, the aqueous humor enters the posterior chamber, flows into the anterior chamber (via the pupil), and is drained from the anterior chamber via the canal of Schlemm.

Obstruction of the canal of Schlemm can lead to which clinical condition?

Glaucoma

What are the consequences of glaucoma?

The resultant increase in intraocular pressure can cause retinal damage and blindness.

What is "vitreous humor"?

Watery fluid posterior to the lens that functions to support the lens and transmit light

Ocular Tunics

What are the 3 ocular tunics?

The fibrous tunic, the vascular tunic, and the retina

Fibrous Tunic

What 2 structures form the fibrous tunic of the eyeball?

The sclera (covers the posterior five sixths of the eyeball) and the cornea (covers the anterior one sixth of the eyeball)

Which structures pierce the sclera?

1. CN II (the optic nerve)
2. The central artery of the retina, a branch of the ophthalmic artery (encased within the dural sheath of CN II)
3. The ciliary nerve, artery, and vein

What is the function of the cornea?

Refraction of light

What clinical condition may result from trauma to the cornea or from an ocular foreign body?

Corneal abrasion. This condition is characterized by eye pain and irritation. It is diagnosed using a slit lamp or by examining the eye with a "Wood's lamp" after instilling fluorescein dye (the abraded area will light up).

What is the name of the site where the cornea and the sclera meet?

The limbus

Vascular Tunic

Which 3 structures comprise the middle vascular tunic of the eyeball?

The choroid, the ciliary body (i.e., the ciliary muscle and the ciliary processes), and the iris

Which structure is responsible for nourishing the retina?

The choroid

What action changes the convexity of the lens?

Contraction of the ciliary muscle

When the lens is focusing on distant objects, which change occurs within it?

It flattens. Flattening of the lens is achieved by relaxation of the ciliary muscle, which leads to contraction of the suspensory ligament. With age, the lens loses flexibility and cannot bend adequately to focus on near objects (presbyopia).

What is the nerve supply to the ciliary muscle?

Parasympathetic fibers from CN III (the oculomotor nerve), via the short ciliary nerves (which, in turn, are postganglionic fibers from the ciliary ganglion)

What is the name of the central pigmented diaphragm in the middle eye layer?

The iris (i.e., the colored part of the eye)

What is the central aperture of the iris called?

The pupil

Which type of fibers innervate:

The sphincter pupillae (i.e., the circular muscle fibers of the iris)?

Parasympathetic fibers from CN III (the oculomotor nerve), via the short ciliary nerves from the ciliary ganglion

The dilator pupillae (i.e., the radial muscle fibers of the iris)?

Sympathetic fibers, via the long ciliary nerves (branches of nasociliary nerve). Sympathetic fibers equal "fight or flight" so associated with dilated pupils.

Retina

What are the 2 layers of the retina?

The pigmented retina and the neural retina. These layers are normally fused, but may separate with trauma (retinal detachment), a condition that mandates emergent treatment to preserve vision.

Where are the photoreceptors (i.e., rods and cones) found?

In the neural retina

Which structures are specialized for vision in dim light?

Rods

Which structures are specialized for visual acuity and color vision?

Cones

The greatest visual acuity is found on which portion of the retina?

The macula (near the center)

What is the name of the central depression in the macula?

The fovea centralis (contains only cones)

Axons from ganglion cells of the retina converge to form which structure?

CN II (the optic nerve)

What is the name of the origin of CN II on the retina?	The optic disc
What is the center of the optic disc called?	The optic cup

Innervation of the Eye

What is the name of the parasympathetic ganglion in the posterior orbit, lateral to the optic nerve?	The ciliary ganglion
The ciliary ganglion transmits parasympathetic fibers to which structures via the short ciliary nerves?	The sphincter pupillae and ciliary muscles
Which nerve provides the sense of sight?	CN II (the optic nerve)
Is CN II part of the central or peripheral nervous system?	The optic nerve is unusual in that it is invested by all 3 layers of meninges (pia mater, arachnoid mater, dura mater) and is therefore part of the *central* nervous system.

What is the clinical significance of this feature of the optic nerve?

Increased intracranial pressure (as with bleeding, tumors) is transmitted to the retina via the subarachnoid space, which contains cerebrospinal fluid (CSF). This leads to characteristic retinal abnormalities that may be noted on ophthalmoscopic examination.

Where does the optic nerve from 1 eye join the optic nerve from the opposite eye?

At the optic chiasm. This chiasm lies in close proximity to the pituitary gland, such that pituitary tumors may present with a characteristic loss of lateral vision in both eyes (bitemporal hemianopsia).

Vasculature of the Eye

The optic disc is pierced by which blood vessel?

The central artery of the retina

What can result from occlusion of the central artery of the retina?

Instant blindness. This is a classic symptom of embolic debris from plaque in the proximal carotid arterial system and is termed amaurosis fugax.

Which feature of the central artery of the retina makes its occlusion such an emergency?

It is an end artery (i.e., the area supplied by the central artery of the retina has no collateral [alternative] circulation).

What is the venous drainage from the orbit?

The superior and inferior ophthalmic veins, draining into the cavernous sinus (i.e., 1 of the venous [dural] sinuses)

Musculature of the Eye

Define the action of each of the following muscles:
    Superior rectus muscle

In Isolation: Moves the eyeball superiorly and medially
In conjunction with the inferior oblique muscle: Moves the eyeball superiorly

    Inferior rectus muscle

In Isolation: Moves the eyeball inferiorly and medially
In conjunction with the superior oblique muscle: Moves the eyeball inferiorly

    Medial rectus muscle

*Ad*ducts the eyeball (pulls it toward the midline)

    Lateral rectus muscle

*Ab*ducts the eyeball (pulls it away from the midline)

    Superior oblique muscle

In isolation: Moves the eyeball inferiorly and laterally
In conjunction with the inferior rectus muscle: Moves the eyeball inferiorly

    Inferior oblique muscle

In isolation: Moves the eyeball superiorly and laterally

In conjunction with the superior rectus muscle: Moves the eyeball superiorly

What is the common origin of all 4 rectus muscles?

They all arise from a common tendinous ring around the optic canal in the posterior orbit.

Where do the rectus muscles insert?

Into the anterior sclera

How do the oblique muscles move the eye straight up and down?

To move the eye straight up or down, the obliques recruit the recti muscles—the inferior oblique works with the superior rectus, and the superior oblique works with the inferior rectus.

Turning the eye medially and looking up and down tests which muscles?

The superior and inferior obliques

Turning the eye laterally and looking up and down tests which muscles?

The superior and inferior recti

What are the origin, course, and insertion of the superior oblique muscle?

From the body of the sphenoid bone, the muscle forms a tendon that runs anteriorly to reach the trochlea, where it turns posteriorly and courses laterally to insert on the sclera inferior to the superior rectus.

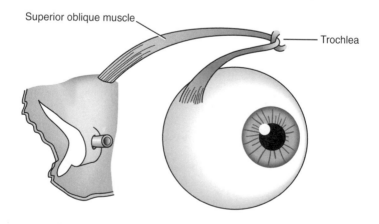

Superior oblique muscle

Trochlea

What are the origin and insertion of the inferior oblique muscle?

Origin: The floor of the orbit
Insertion: The sclera, inferior to the lateral rectus

Describe the innervation of the extraocular muscles.

CN III (*oculomotor* nerve) innervates the superior rectus, inferior rectus, medial rectus, and inferior oblique muscles.
CN IV (*trochlear* nerve) innervates the superior oblique muscle (which forms a sling around the *trochlea*).
CN VI (*abducens* nerve) innervates the lateral rectus muscle (which *abducts* the eye, pulling it laterally).

 A lesion of CN III will lead to what deficit?

Limitation of ocular elevation, depression, and adduction. At rest, the eye will be externally deviated (held in position by the unopposed contraction of the lateral rectus muscle). There will also be ptosis (eyelid droop from loss of innervation to levator palpebrae superioris) and a widely dilated pupil (loss of parasympathetics to sphincter).

The *superior* division of CN III innervates which structures?

Superior rectus and levator palpebrae superioris muscles

The *inferior* division of CN III innervates which structures?

Medial and inferior recti, inferior oblique and ciliary muscles, and sphincter pupillae

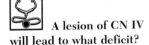 **A lesion of CN IV will lead to what deficit?**

Failure of eye to depress fully in adduction. At rest, there is slight upward deviation of the involved eye.

A lesion of CN VI will lead to what deficit?

Failure of eye to abduct

EYELIDS

What is the name of the longitudinal opening between the upper and lower eyelids?

The palpebral fissure

What is the name of the mucous membrane lining the inner eyelids and anterior surface of the eyeball?	The conjunctiva
Which muscle closes the eyelid?	The orbicularis oculi
Which muscle opens the eyelid?	The levator palpebrae superioris (assisted by the superior tarsal muscle)
What is the origin and insertion of the levator palpebrae superioris muscle?	**Origin:** The lesser wing of the sphenoid bone **Insertion:** The skin of the upper eyelid
Which nerve innervates the levator palpebrae superioris muscle?	The superior division of CN III (the oculomotor nerve)

LACRIMAL APPARATUS

Tears are produced by which gland?	The lacrimal gland
Trace the flow of tears from the lacrimal gland to the nose.	Tears from the lacrimal gland enter the eye at the upper orbit, circulate across the cornea, gather in the lacus lacrimalis, and enter the lacrimal canaliculi, which open into the lacrimal sac. The nasolacrimal duct then empties into the inferior meatus of the nose.
What is the name of the opening into the lacrimal canaliculi, located in the medial eye?	The punctum lacrimalis
Which structure passes through the nasolacrimal canal?	The nasolacrimal duct
Describe the innervation of the lacrimal gland?	Preganglionic parasympathetic fibers from CN VII (the facial nerve) travel from the lacrimal nucleus to the pterygopalatine ganglion, and then postganglionic fibers join CN V_2 (the maxillary division of the

trigeminal nerve). Within CN V$_2$, fibers travel in the zygomatic nerve, and then cross over to the lacrimal nerve and on to the lacrimal gland. Note: Fibers travel with CN V$_2$ but originate with CN VII.

In addition to the lacrimal gland, what other structures are innervated by the lacrimal nerve?

The conjunctiva and the skin of the upper eyelid

THE EAR

What are the 3 general parts of the ear?

The external ear, the middle ear (tympanic cavity), and the inner ear (labyrinth)

The sensory organs for hearing and balance lie within which part?

The inner ear

EXTERNAL EAR

Which 2 structures make up the external ear?

The auricle and external auditory meatus

What is the function of the auricle?

It collects sound vibrations.

What is the function of the external auditory (acoustic) meatus?

It leads into the external auditory canal, which transmits sound from the auricle to the tympanic membrane. Inflammation of the external auditory canal (otitis externa) leads to ear pain. As this condition commonly occurs after water enters the canal, it is known as "swimmer's ear."

Which nerves innervate the auricle?

1. The auricular nerve, a branch of CN X (the vagus nerve)
2. The greater auricular and lesser occipital nerves (branches from the cervical plexus)
3. The auriculotemporal nerve, a branch of CN V$_3$ (the mandibular branch of the trigeminal nerve)

Which portion of the external auditory meatus is formed from cartilage?

The external third

Which nerves innervate the external auditory meatus?

The auriculotemporal and auricular nerves

Which 3 arteries supply the auricle and external auditory meatus?

1. The posterior auricular artery (a branch of the external carotid artery)
2. The deep auricular artery (a branch of the maxillary artery)
3. The auricular branch of the superficial temporal artery

What are the 2 means of venous drainage from the auricle and external auditory meatus?

The external jugular and maxillary veins, and the pterygoid plexus

Which structure lies at the end of the external auditory meatus, marking the medial boundary of the external ear?

The tympanic membrane ("eardrum"). This structure is readily visible through an otoscope and may appear inflamed in the setting of middle ear infection (otitis media). In severe cases, the tympanic membrane may perforate.

Label the structures on the following figure of the tympanic membrane:

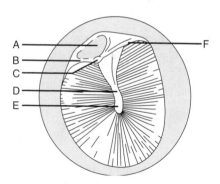

A = Flaccid part
B = Anterior mallear fold
C = Lateral process of the malleus
D = Handle of the malleus
E = Umbo
F = Posterior mallear fold

How many layers make up the tympanic membrane?	3 (2 epithelial layers and an intermediate fibrous layer)
Which nerves innervate the:	
Outer surface of the tympanic membrane?	The auriculotemporal and auricular nerves
Inner (medial) surface of the tympanic membrane?	The tympanic branch of CN IX (the glossopharyngeal nerve)

MIDDLE EAR (TYMPANIC CAVITY)

The middle ear is a space within which bone?	The petrous part of the temporal bone
What are the boundaries of the middle ear:	
Superiorly (i.e., the roof)?	The tegmen tympani (a thin plate of bone that is part of the petrous temporal bone)
Inferiorly (i.e., the floor)?	The jugular fossa and a thin plate of bone
Anteriorly?	The carotid canal
Posteriorly?	Mastoid air cells and the mastoid antrum. Because of their proximity to the middle ear, mastoid air cell involvement commonly complicated otitis media in the days before good antibiotics.
Laterally?	The tympanic membrane
Medially?	The lateral wall of the inner ear
An opening in the anterior wall of the middle ear exists for which structure?	The auditory (eustachian) tube
The auditory (eustachian) tube connects the middle ear with which space?	The nasopharynx. This represents a potential avenue for the spread of resident bacteria from the nasopharynx into the middle ear (normally sterile).

Contraction of which 2 muscles opens the auditory (eustachian) tube?

The tensor veli palatini and salpingopharyngeus muscles

Which 3 nerves innervate the middle ear?

1. The auriculotemporal nerve
2. The tympanic branch of CN IX (the glossopharyngeal nerve)
3. The auricular nerve

Which 3 bones lie in the middle ear?

The malleus, incus, and stapes (the 3 ossicles, sometimes referred to as "the hammer, anvil, and stirrup," respectively, because of their shapes)

Which ossicle attaches to the tympanic membrane?

The malleus

Name the 2 openings on the lateral wall of the inner ear.

The oval window and the round window

Upon which opening does the footplate of the stapes rest?

The oval window

Name the 2 muscles in the tympanic cavity.

The stapedius and tensor tympani muscles

What is the origin and insertion of the:
  Stapedius muscle?

Origin: The posterior wall of the tympanum (pyramidal eminence)
Insertion: The neck of the stapes

  Tensor tympani muscle?

Origin: The cartilaginous portion of the auditory (eustachian) tube
Insertion: The handle of the malleus

What is the claim to fame of the stapedius muscle?

It is the smallest skeletal muscle in the human body.

What is the function of the stapedius muscle?

Its contraction prevents loud noises from injuring the inner ear.

Paralysis of the stapedius muscle leads to "hyperacusis" (i.e., the abnormal acuteness of hearing).

What is the function of the tensor tympani muscle?

It dampens vibrations of the malleus (although some argue that this muscle has little physiologic function).

Which nerve innervates the stapedius muscle?

CN VII

Which nerve innervates the tensor tympani muscle?

CN V_3 (mandibular branch of trigeminal nerve)

Which 3 arteries supply the middle ear?

1. The posterior auricular artery
2. The anterior tympanic artery
3. The caroticotympanic artery

INNER EAR (LABYRINTH)

Which bone houses the inner ear?

The petrous portion of the temporal bone

Which 3 structures make up the bony labyrinth?

1. The vestibule
2. The cochlea
3. The semicircular canals

Which structures comprise the membranous labyrinth, a series of communicating sacs and ducts that is suspended in the bony labyrinth?

1. The utricle and saccule (housed in the vestibule
2. The cochlear duct
3. The semicircular ducts

Which type of fluid is found within the bony labyrinth, surrounding the membranous labyrinth?

Perilymph

Which type of fluid fills the membranous labyrinth?

Endolymph

What is the arterial supply to the inner ear?

The labyrinthine (internal acoustic) artery, a branch of the basilar artery

Identify the labeled
structures on the following
illustration of the middle
and inner ear:

A = Semicircular canal and duct
B = Malleus
C = Incus
D = Stapes
E = External auditory meatus
F = Tympanic membrane
G = Cochlea and cochlear duct
H = Tympanic cavity
I = Eustachian tube

Cochlear Apparatus

After vibrations from the
external ear cause the
tympanic membrane to
vibrate, how are impulses
conveyed to the cochlea
(i.e., the organ of hearing)?

Excitation of the tympanic membrane
causes the ossicles to move. The stapes
(i.e., the final ossicle) then transmits
vibrations to the scala vestibuli via the oval
window. Hair cells along the cochlear
membrane detect the vibratory movement
of that membrane and transduce
vibrations into nerve impulses.

Name the 2 compartments
of the cochlea.

The scala vestibuli (above) and the scala
tympani (below)

Which structure divides the cochlea into the scala vestibuli and the scala tympani?

The spiral lamina

The cochlear duct, located between the scala vestibuli and the scala tympani, contains which sensory organ?

The spiral organ of Corti

Identify the labeled structures on the following figure of the cochlear apparatus:

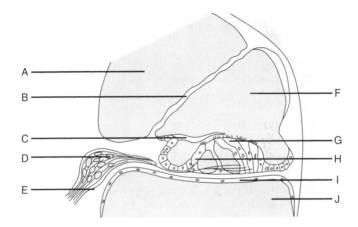

The spiral organ of Corti
A = Scala vestibuli
B = Vestibular membrane
C = Tectorial membrane
D = Cochlear ganglion
E = Cochlear nerve
F = Cochlear duct
G = Outer hair cells
H = Inner hair cells
I = Basilar membrane
J = Scala tympani

Where in the cochlea is the organ of Corti stimulated by:

    Low-frequency sound waves? — Near the apex

    High-frequency sound waves? — Near the base

The scala vestibuli and the scala tympani communicate at the tip of the cochlea through which structure? — The helicotrema

Semicircular Canals and Ducts

What is the function of the semicircular canals of the inner ear? — They sense the *angular* acceleration of the head, thereby regulating balance.

Name the 3 semicircular canals. — Anterior (superior), posterior, and lateral

Which sensory organs are within these canals? — Ampullae

Vestibule

What is the function of the utricle and the saccule? — They are organs that detect *linear* movement, also contributing to balance.

Which sensory organs are found within the utricle and saccule? — Maculae

POWER REVIEW

SKULL

Name the 4 major sutures of the skull.	1. Coronal suture 2. Sagittal suture 3. Squamous suture 4. Lambdoid suture
Name the intersection of the:	
Lambdoid and coronal sutures	Lambda
Sagittal and coronal sutures	Bregma
What are the shapes of the 2 major fontanelles?	**Anterior:** Diamond-shaped **Posterior:** Triangular
For each of the following cranial openings, name the bone where it is located, the structures it contains, and the structures it connects:	
Optic canal	**Location:** Lesser wing of the sphenoid bone **Structures contained:** CN II (the optic nerve), the ophthalmic artery, and the central retinal vein **Structures connected:** The orbit and the middle cranial fossa
Superior orbital fissure	**Location:** Between the greater and lesser wings of the sphenoid bone **Structures contained:** CN III (the oculomotor nerve), CN IV (the trochlear nerve), CN V_1 (the ophthalmic division of the trigeminal nerve), CN VI (the abducens nerve), and the superior ophthalmic vein **Structures connected:** The middle cranial fossa and the orbit

Foramen rotundum

Location: Greater wing of the sphenoid bone

Structures contained: CN V_2 (the maxillary division of the trigeminal nerve)

Structures connected: The middle cranial fossa and the pterygopalatine fossa

Foramen ovale

Location: Greater wing of the sphenoid bone

Structures contained: CN V_3 (the mandibular division of the trigeminal nerve) and the accessory meningeal artery

Structures connected: The middle cranial fossa and the infratemporal fossa

Foramen spinosum

Location: Greater wing of the sphenoid bone

Structures contained: The middle meningeal artery

Structures connected: The middle cranial fossa and the infratemporal fossa

Foramen lacerum

Location: Between the sphenoid bone, occipital bone, and the petrous part of the temporal bone

Structures contained: Greater petrosal nerve and deep petrosal nerve

Structures connected: The middle and posterior cranial fossae and the neck

Which structures pass through the carotid canal?

The internal carotid artery and sympathetic nerves

Which structures pass through the foramen magnum?

The medulla oblongata, CN XI (the spinal accessory nerve), the vertebral arteries, the venous plexus, and the anterior and posterior spinal arteries

Which structures pass through the internal auditory meatus?

CN VII (the facial nerve) and CN VIII (the vestibulocochlear nerve)

Which nerve passes through the cribriform plate of the ethmoid bone?	CN I (the olfactory nerve)
Which skull bones have sinuses?	The frontal, maxillary, ethmoid, and sphenoid bones

SCALP AND SUPERFICIAL AND DEEP FACE

What are the 5 layers of the scalp?	**SCALP** **S**kin **C**onnective tissue **A**poneurosis **L**oose connective tissue **P**eriosteum
What nerve innervates the muscles of facial expression?	CN VII (the facial nerve)
Name the 5 terminal branches of the facial nerve.	Temporal, zygomatic, buccal, mandibular, and cervical
Name the 4 muscles of mastication.	1. Masseter 2. Temporalis 3. Medial pterygoid 4. Lateral pterygoid
Which nerve innervates the muscles of mastication?	CN V_3 (the mandibular division of the trigeminal nerve)
Sensory innervation to the skin of the face is provided by which cranial nerve?	CN V (the trigeminal nerve)
Which nerves provide for sensory innervation to the nose?	**Smell:** CN I (the olfactory nerve) **General sensation:** CN V_1 and CN V_2 (the ophthalmic and maxillary divisions of the trigeminal nerve)
Name the 8 branches of the external carotid artery, from proximal to distal.	1. Superior thyroid artery 2. Ascending pharyngeal artery 3. Lingual artery 4. Facial artery 5. Occipital artery 6. Posterior auricular artery

7. Maxillary artery
8. Superficial temporal artery

ORAL CAVITY

Which 4 muscles comprise the extrinsic musculature of the tongue?	1. Genioglossus 2. Hyoglossus 3. Styloglossus 4. Palatoglossus
Which tongue muscle is NOT innervated by CN XII (the hypoglossal nerve)?	The palatoglossus muscle (innervated by CN X, the vagus nerve)
Lesions of CN XII cause the tongue to deviate to which side?	The side on which the lesion is located
Describe the sensory innervation of the:	
Anterior tongue	**Sensation:** The lingual nerve (a branch of CN V_3, the mandibular division of the trigeminal nerve) **Taste:** The chorda tympani, a branch of CN VII (the facial nerve)
Posterior tongue	CN IX (the glossopharyngeal nerve) and CN X (the vagus nerve), for both sensation and taste

NOSE

Where is each paranasal sinus ostia located:	
Maxillary?	Middle meatus
Anterior ethmoid?	Middle meatus
Posterior ethmoid?	Superior meatus
Frontal?	Middle meatus
Sphenoid?	Sphenoethmoid recess
Where is the lacrimal gland ostium located?	Inferior meatus

| Which structures contribute to formation of the nasal septum? | Quadrangular cartilage, perpendicular plate of the ethmoid bone, vomer, maxillary crest, and palatine bone |

SALIVARY GLANDS

| What are the 3 major sets of salivary glands? | Parotid, submandibular, and sublingual |

| Where does the parotid duct enter the mouth? | Adjacent to the 2nd upper molar |

| Where does the submandibular duct enter the mouth? | Inferior to the tongue, adjacent to the frenulum |

| What 3 major structures traverse the parotid gland? | 1. CN VII (facial nerve)
2. Retromandibular vein
3. External carotid artery |

PHARYNX

| What are the 2 major groups of pharyngeal muscles? | 1. The pharyngeal constrictor (external) muscles, consisting of the superior constrictor, the middle constrictor, and the inferior constrictor
2. The longitudinal (internal) muscles, consisting of the palatopharyngeus, salpingopharyngeus, and stylopharyngeus muscles |

| What are the muscles of the soft palate? | 1. Tensor veli palatini
2. Levator veli palatini
3. Palatoglossus
4. Palatopharyngeus
5. Musculus uvulae |

| Which of these muscles is not innervated by CN X (the vagus nerve)? | The tensor veli palatini (innervated by CN V_3, the mandibular division of the trigeminal nerve) |

Which nerve provides sensory innervation to the:	
Nasopharynx?	CN V_2 (the maxillary division of the trigeminal nerve)
Oropharynx?	CN IX (the glossopharyngeal nerve)

Laryngopharynx?	CN X (the vagus nerve)
The palatine tonsils lie between which 2 arches (folds)?	The palatoglossal and palatopharyngeal arches, which form the boundary of the oropharynx
Which structure connects the tympanic cavity with the lateral wall of the nasopharynx?	The auditory (eustachian) tube

EYE

CN III (the oculomotor nerve) innervates which muscles?	**Superior division:** The superior rectus and levator palpebrae superioris muscles **Inferior division:** The medial and inferior recti and the inferior oblique muscles; in addition, the inferior division of CN III provides parasympathetic fibers to the ciliary muscle and sphincter pupillae
CN IV (the trochlear nerve) innervates what muscle?	The superior oblique muscle
CN VI (the abducens nerve) innervates which muscle?	The lateral rectus muscle
Describe the innervation of the:	
Sphincter pupillae	Parasympathetic fibers (via CN III, the ciliary ganglion, and the short ciliary nerves)
Dilator pupillae	Sympathetic fibers via the long ciliary nerve
Which structures are specialized for:	
Vision in dim light?	Rods
Color vision?	Cones

EAR

Which 3 bones are contained in the middle ear?

1. The malleus (attaches to the tympanic membrane, the link to the external ear)
2. The incus
3. The stapes (attaches to the oval window of the cochlea, the link to the inner ear)

Remember: The round window is the end of the cochlea.

Name the 2 muscles of the inner ear and the nerves that innervate them.

1. The stapedius muscle (innervated by CN VII)
2. The tensor tympani (innervated by CN V_3)

Which 3 structures make up the bony labyrinth of the inner ear?

The vestibule, the semicircular canals, and the cochlea

What is the function of the vestibule?

Aids in balance through detection of linear movement

What is the function of the semicircular canals?

Regulate balance by sensing angular acceleration of the head

What is the function of the cochlea?

Hearing

Which fluid fills the bony labyrinth?

Perilymph

Which sensory organs are found within the utricle and saccule?

Maculae

What is the arterial supply to the inner ear?

The labyrinthine artery (a branch of the basilar artery)

3   The Central Nervous System

INTRODUCTION

What are the 3 components of a neuron?

1. Cell body
2. Dendrites (carry impulses to the cell body)
3. Axon (carries impulses away from the cell body)

What is the difference between a ganglion and a nucleus?

Ganglion: A collection of cell bodies outside of the central nervous system (CNS)

Nucleus: A collection of cell bodies within the CNS

What is the difference between gray matter and white matter?

Gray matter (cell bodies) is unmyelinated; white matter (axons) is myelinated.

Which structures comprise the CNS?

The brain and the spinal cord

How many spinal nerves and cranial nerves leave the CNS?

31 spinal nerves and 12 cranial nerves (Think, "Eat 31 flavors in 12 months.")

THE BRAIN

Cerebrum

What is the cerebrum?

The largest part of the brain, concerned mostly with higher-order thinking

Where in the cranium is the cerebrum located?

The anterior and middle cranial fossae

How is the cerebrum divided?

The cerebrum is divided into right and left **hemispheres**. Each hemisphere contains 4 **lobes**. Each lobe is divided into a number of **gyri** (folds). Each gyrus contains organized collections of **neurons**.

What external landmark separates the cerebral hemispheres from one another?

The longitudinal (interhemispheric) fissure

What are gyri and sulci?

Gyri are elevated folds on the surface of the cerebral hemispheres. Sulci are the grooves that separate the gyri from each other. This architecture increases the surface area of the brain.

Identify the 4 lobes of the brain on the following figure:

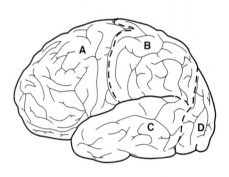

A = Frontal lobe
B = Parietal lobe
C = Temporal lobe
D = Occipital lobe

What separates the:
    Frontal lobe from the parietal lobe?

The central sulcus

    Parietal lobe from the occipital lobe?

The parietooccipital sulcus

    Temporal lobe from the frontal and parietal lobes?

The lateral sulcus (i.e., the sylvian fissure)

On the cerebral cortex, which region is concerned with:

Sensory information?

The **postcentral gyrus** (i.e., the part of the parietal lobe immediately posterior to the central sulcus)

Motor signals?

The **precentral gyrus** (i.e., the part of the frontal lobe immediately anterior to the central sulcus)

What is the general function of the frontal lobe of the cerebral cortex?

The frontal lobe mediates higher cerebral function/thought. Damage to the frontal lobe from a tumor or stroke may lead to changes in personality and behavior.

Where is the superior temporal gyrus located, and what is its role?

The superior temporal gyrus is located in the temporal lobe immediately below the sylvian fissure. It is concerned with auditory stimuli.

Where is the visual cortex located?

In the occipital lobe. Accordingly, an infarction (stroke) in the occipital lobe generally results in visual field loss.

Which temporal lobe structure is involved with memory?

The hippocampus

What is the cerebrum composed of?

Gray matter (cell bodies, superficial) and white matter (axonal tracts, deep)

What major white matter structure connects the hemispheres?

The corpus callosum (hemispheres also connected by the "anterior commissure") Damage to or congenital absence(agenesis) of the corpus callosum may lead to subtle deficits in memory, judgment, and behavior related to a lack of coordination between the 2 cerebral hemispheres ("split-brain syndrome").

What are the basal ganglia?

Deep gray matter—specific nuclei positioned deep in the base of the cerebral hemispheres that are involved in motor function

Which 5 nuclei comprise the basal ganglia?

1. The caudate nucleus
2. The putamen
3. The globus pallidus
4. The substantia nigra
5. The subthalamic nucleus

Which 3 main arteries supply blood to the cerebrum?

1. Anterior cerebral artery
2. Middle cerebral artery
3. Posterior cerebral artery

Cerebellum

Where in the cranium is the cerebellum located?

In the posterior cranial fossa

What is the main role of the cerebellum?

Coordination of movement and postural adjustment

What clinical manifestations are characteristic in patients with cerebellar lesions?

Unsteady gait, poor coordination of movement, tremor of the hands

Which 3 arteries supply the cerebellum?

1. The posterior inferior cerebellar artery (a branch of the vertebral artery)
2. The anterior inferior cerebellar artery (a branch of the basilar artery)
3. The superior cerebellar artery (a branch of the basilar artery)

Diencephalon

Which major structures are part of the diencephalon?

The thalamus and hypothalamus

Which structure is considered the "relay station" of the brain?

The thalamus

Which physiologic system is the hypothalamus involved with?	The endocrine system. The hypothalamus is involved in the release of hormones to the pituitary gland.

Brainstem

What are the 3 major parts of the brainstem?	The midbrain (mesencephalon), pons, and medulla oblongata
Which part of the brainstem mediates visual/auditory reflexes and maintains consciousness?	The midbrain
What is the function of the medulla oblongata?	It carries tracts to and from the spinal cord and contains "autonomic" centers that mediate cardiovascular, respiratory, and gastrointestinal function.
What is the function of the pons?	In addition to containing additional "autonomic" centers, the pons has a role in regulating hearing, balance, and facial movements.

MENINGES OF THE BRAIN

What are meninges?	Protective coverings ("membranes") over the surface of the brain
What are the 3 meningeal layers, from superficial to deep?	1. Dura mater 2. Arachnoid mater 3. Pia mater
What separates the dura mater from the arachnoid mater?	The subdural space (a potential space)
What causes a subdural hematoma?	Traumatic rupture of the cerebral veins as they pass from the brain surface into the venous (dural) sinuses
What is the space between the arachnoid mater and the pia mater referred to as?	The subarachnoid space

How is the subarachnoid space different from the subdural space?

Unlike the subdural space, the subarachnoid space is a true space. It contains cerebrospinal fluid (CSF), which cushions the brain.

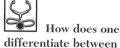 **How does one differentiate between a *sub*dural and *epi*dural hematoma on a computed tomography (CT) scan?**

A subdural hematoma usually takes on a crescent shape, whereas an epidural hematoma has a lenticular (lens-like) shape.

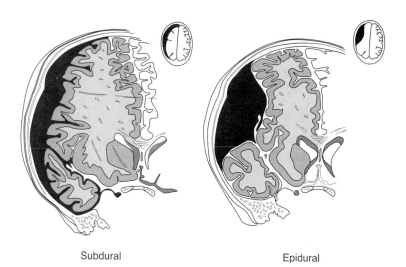

Subdural   Epidural

Dura Mater

What artery provides the major blood supply to the dura mater?

The middle meningeal artery, a branch of the maxillary artery

Describe the course of the middle meningeal artery.

It branches from the maxillary artery in the infratemporal fossa, enters the cranium via foramen spinosum, and then runs in a groove on the inner aspect of the temporal bone.

What clinical entity results from rupture of the middle meningeal artery?

Epidural hematoma (bleeding into potential space between dura and overlying skull)

What are the 4 dural folds?

1. The falx cerebri
2. The tentorium cerebelli
3. The falx cerebelli
4. The diaphragma sellae

What is the anatomic function of the dural folds?

They separate the cranial cavity into compartments.

Falx Cerebri

Where is the falx cerebri located?

In the longitudinal fissure (i.e., the fissure that separates the cerebral hemispheres from one another)

What are the anterior and posterior attachments of the falx cerebri?

Anterior: The crista galli of the ethmoid bone and the frontal crest of the frontal bone
Posterior: Internal occipital protuberance

Which structures are found in the inferior and superior margins of the falx cerebri?

The inferior and superior sagittal sinuses, respectively

Tentorium Cerebelli

Where is the tentorium cerebelli located?

The tentorium cerebelli separates the cerebrum from the brainstem and cerebellum. It lies horizontally between the occipital lobes of the cerebral hemispheres and the cerebellum. Structures superior and inferior to this membrane are referred to as "supratentorial" and "infratentorial," respectively.

What are the attachment points for the tentorium cerebelli:
    Anteriorly?

The clinoid processes

    Laterally?

The temporal and parietal bones

    Posteriorly?

The occipital bone

Which dural structure is continuous with the tentorium cerebelli?

The falx cerebri

Falx Cerebelli

Where is the falx cerebelli located?

In the longitudinal fissure between the 2 cerebellar hemispheres (Hint: The 2 falx structures both run between hemispheres.)

Diaphragma Sellae

What structure does the diaphragma sellae form a "roof" over?

The pituitary gland, within the hypophyseal fossa in the sella turcica

A small opening in the center of the diaphragma sellae transmits which structure?

The pituitary stalk (infundibulum), which connects the pituitary with the hypothalamus. Head trauma may occasionally result in transection of this stalk, with subsequent pituitary dysfunction (hypopituitarism).

Where does the diaphragma sellae attach?

To the clinoid processes

What important structure lies on top of the diaphragma sellae?

The optic chiasm. The chiasm's proximity to the sella turcica makes it vulnerable to compression from pituitary tumors.

ARACHNOID MATER

What does the term "arachnoid" translate to?

"Spiderlike." The arachnoid layer is a delicate, cobweblike structure.

Pia Mater

What is contained within the pia mater?

Blood vessels, blood vessels, and more blood vessels (highly vascular)

VASCULATURE OF THE BRAIN

Arteries

Which 2 sets of arteries constitute the blood supply of the brain?

The internal carotid arteries ("anterior circulation") and the vertebral arteries ("posterior circulation"). Infarcts in the anterior circulation are often the result of embolic debris from atherosclerotic plaque in the proximal internal carotid artery. Removal of this plaque with the associated arterial intima and media is termed carotid endarterectomy.

Internal Carotid Arteries

Describe the course of the internal carotid artery.

After arising from the common carotid artery in the neck, the internal carotid artery enters the skull through the carotid canal (located in the petrous portion of the temporal bone), passes through the cavernous sinus, and enters the subarachnoid space.

Name the 4 parts of the internal carotid artery.

1. Cervical part
2. Petrous part
3. Cavernous part
4. Cerebral part

Does the internal carotid artery give off any branches in the neck before entering the cranium?

No

What are the 3 major intracranial branches of the internal carotid artery?

1. Ophthalmic artery
2. Posterior communicating artery
3. Anterior choroid artery

What are the 2 terminal branches of the internal carotid artery?

Middle cerebral and anterior cerebral arteries

Which branch of the internal carotid artery joins the posterior cerebral artery?

The posterior communicating artery

Which is larger, the middle cerebral artery or the anterior cerebral artery?

The middle cerebral artery

Vertebral Arteries

Describe the course of the vertebral arteries.

They arise from the first part of the subclavian artery, ascend through the transverse foramina of vertebrae C1–C6, and then access the intracranial space via the foramen magnum. The segment of the vertebral artery within the transverse foramina is susceptible to injury from spine fractures.

What are the 3 major branches of the vertebral artery?

1. Anterior spinal artery
2. Posterior spinal artery
3. Posterior inferior cerebellar artery

Which of these branches is the largest?

The posterior inferior cerebellar artery

The vertebral arteries join on the surface of the pons to form which structure?

The basilar artery

What structures are supplied by the basilar artery?

Part of the cerebellum and the pons. Unlike infarcts of the anterior circulation, which are most often embolic in origin, infarcts in the vertebrobasilar distribution are usually related to hypertension.

What are the 4 major branches of the basilar artery?

1. Pontine branches
2. Anterior inferior cerebellar artery
3. Labyrinthine artery
4. Superior cerebellar artery

The basilar artery ends by dividing into which vessels?

The left and right posterior cerebral arteries

What is a berry aneurysm?

A congenital dilatation of a blood vessel within the brain

What is the clinical significance of a berry aneurysm?

It may rupture, leading to subarachnoid hemorrhage, or it may cause symptoms by enlarging and impinging on other structures.

Circle of Willis

What is the circle of Willis?

An arterial anastomotic network on the base of the brain, formed by the communication of the right and left internal carotid arteries and the right and left vertebral arteries

Identify the arteries on the following drawing of the circle of Willis:

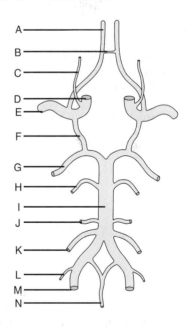

A
B
C
D
E
F
G
H
I
J
K
L
M
N

A = Anterior cerebral artery
B = Anterior communicating artery
C = Ophthalmic artery
D = Internal carotid artery
E = Middle cerebral artery
F = Posterior communicating artery
G = Posterior cerebral artery
H = Superior cerebellar artery
I = Basilar artery
J = Labyrinthine artery
K = Anterior inferior cerebellar artery
L = Posterior inferior cerebellar artery
M = Vertebral artery
N = Anterior spinal artery

Cerebral Veins and Venous (Dural) Sinuses

What is unique about cerebral veins?	They are valveless.
What is the origin of the great cerebral vein (of Galen)?	It is formed by the union of the 2 internal cerebral veins.
What are venous sinuses?	Channels within the dura mater that route venous blood and CSF from the brain to the systemic venous circulation
Which veins form an anastomotic network connecting the scalp and the venous sinuses?	The diploic veins
Which veins directly connect the venous sinuses with the scalp?	The emissary veins. Tearing of these veins with head trauma may lead to a subdural hematoma.
What vein do the venous sinuses ultimately drain into?	The internal jugular vein
Which sinus lies within the convex (superior) border of the falx cerebri?	The superior sagittal sinus
Which veins empty into the superior sagittal sinus?	1. The diploic veins 2. The meningeal veins 3. The emissary veins

**Label the venous sinuses on
the following figure:**

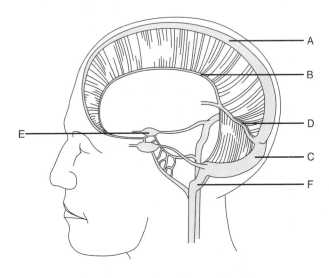

A = Superior sagittal sinus
B = Inferior sagittal sinus
C = Transverse sinus
D = Straight sinus
E = Cavernous sinus
F = Sigmoid sinus

The superior sagittal sinus ends by becoming continuous with which other sinus?	The transverse sinus
Which sinus, located in the free (inferior) edge of the falx cerebri, is joined by the great cerebral vein (of Galen)?	The inferior sagittal sinus
What is the course of blood from the inferior sagittal sinus to the inferior jugular vein?	Inferior sagittal sinus, straight sinus, transverse sinus, sigmoid sinus, internal jugular vein

Which sinus is formed by the junction of the inferior sagittal sinus and the great cerebral vein (of Galen)?

The straight sinus

In between which dural folds does the straight sinus run?

In the line of attachment between the falx cerebri and the tentorium cerebelli

Into which structure does the straight sinus empty?

The transverse sinus

Where is the transverse sinus located?

The transverse sinus occupies the attached margin of the tentorium cerebelli.

After leaving the tentorium cerebelli, what does the transverse sinus become?

The sigmoid sinus

Describe the course of the sigmoid sinus.

The sigmoid sinus runs through a groove in the mastoid part of the temporal bone. It then runs anteriorly and inferiorly to reach the jugular foramen, where it becomes continuous with the internal jugular vein.

Which sinus is found on either side of the sella turcica?

The cavernous sinus. Infections of the face and orbit may spread via venous channels to this sinus, leading to cavernous sinus thrombosis. This condition is associated with headaches and deficits in extraocular movements (due to involvement of nearby cranial nerves (CN) III–VI).

Which structures provide communication between the 2 sides of the cavernous sinus?

Intercavernous sinuses

Which 4 cranial nerves are found within the lateral wall of the cavernous sinus?

1. CN III (the oculomotor nerve)
2. CN IV (the trochlear nerve)
3. CN V_1 (the ophthalmic division of the trigeminal nerve)
4. CN V_2 (the maxillary division of the trigeminal nerve)

Which nerve and which vessel run inside the cavernous sinus?

1. CN VI (the abducens nerve)
2. The ophthalmic artery, a branch of the internal carotid artery

Which 3 veins drain into the cavernous sinus?

The superior and inferior ophthalmic veins, and the great cerebral vein (of Galen)

Which small sinus, lying in the margin of the tentorium cerebelli, runs from the cavernous sinus to the transverse sinus?

The superior petrosal sinus

Which small sinus drains the cavernous sinus into the internal jugular vein?

The inferior petrosal sinus

Describe the 2 courses venous blood may take between the cavernous sinus and the internal jugular vein.

1. Cavernous sinus, superior petrosal sinus, transverse sinus, sigmoid sinus, internal jugular vein
2. Cavernous sinus, inferior petrosal sinus, internal jugular vein

Which is the smallest of the venous sinuses?

The occipital sinus

Where is the occipital sinus located?

In the attached margin of the falx cerebelli

Where does the occipital sinus drain?

Into the "confluence of sinuses," a point in the occipital area where 5 of the 6 major venous sinuses converge

CEREBROSPINAL FLUID

What is CSF?

CSF is the liquid that fills the subarachnoid space, surrounding and cushioning the brain and spinal cord. Lumbar puncture ("spinal tap") involves needle aspiration of CSF from the subarachnoid space in the lumbar spine and is most commonly performed to rule out infection (meningitis).

What is the source of CSF?

CSF is formed by the choroid plexus, within the cerebral ventricles.

Name the 4 ventricles.

1. Right lateral ventricle
2. Left lateral ventricle
3. Third ventricle
4. Fourth ventricle

What is the name of the thin wall that separates the right and left lateral ventricles?

The septum pellucidum

Which ventricle is contained within the diencephalon?

The third ventricle

Which ventricle lies between the pons and the cerebellum?

The fourth ventricle

Describe the course of CSF from the lateral ventricle to the fourth ventricle.

Lateral ventricle, interventricular foramen (of Monro), third ventricle, cerebral aqueduct (of Sylvius), fourth ventricle

How does CSF travel from the fourth ventricle to the subarachnoid space?

From the fourth ventricle, CSF passes through the foramen of **M**agendie (**m**edial) and the 2 foramina of **L**uschka (**l**ateral) into "cisterns" within the subarachnoid space.

How is CSF reabsorbed from the subarachnoid space into the venous circulation?

Through the arachnoid villi, projections of the arachnoid mater that drain into the venous (dural) sinuses.Obstruction of CSF flow or uptake results in hydrocephalus, a condition associated with dilated ventricles and secondary pressure on vital intracranial structures.

What are aggregations of arachnoid villi called?

Arachnoid granulations

Where does most CSF pass into the venous blood?

At the superior sagittal sinus

THE SPINAL CORD

VERTEBRAL COLUMN

What structure supports and protects the spinal cord?	The vertebral column, composed of vertebrae and intervertebral disks
What is a vertebral foramen?	The large central circular opening within each vertebra, bounded by lamina posteriorly, pedicles laterally, and the vertebral body anteriorly. Collectively, these openings form the vertebral canal, which houses the spinal cord and its associated structures.
How many vertebrae comprise the vertebral column?	33, although only 24 are movable in adults
What are the 5 regions of the vertebral column, and how many vertebrae comprise each?	**Cervical:** 7 **Thoracic:** 12 **Lumbar:** 5 **Sacral:** 5 (fused) **Coccygeal:** 4 (fused)

Where does the spinal cord end in:

Adults?	Vertebral level L2
Newborns?	Vertebral level L3
What is the tapered (cone-shaped) lower end of the spinal cord called?	The conus medullaris

SPINAL NERVES

What is the distribution of the 31 pairs of spinal nerves that leave the spinal cord via the intervertebral foramina?	**Cervical:** 8 **Thoracic:** 12 **Lumbar:** 5 **Sacral:** 5 **Coccygeal:** 1
Where does spinal nerve C5 exit the vertebral column relative to vertebra C5?	Above it

Where does spinal nerve T5 exit the vertebral column relative to vertebra T5?

Below it

Explain why spinal nerve C5 exits above its associated vertebra, while spinal nerve T5 exits below.

This is a function of the fact that there are 8 cervical spinal nerves and only 7 cervical vertebra; spinal nerve C8 exits below vertebra C7 and above vertebra T1, forcing spinal nerve T1 to exit below vertebra T1, and so on.

What is the cauda equina ("horse's tail")?

The splayed bundle of elongated spinal nerve roots caudal to the termination of the spinal cord (i.e., below vertebra L2). Its compression or injury may result in cauda equina syndrome, characterized by sensory disturbances in a "saddle" distribution and bladder/bowel dysfunction.

How are the spinal nerves formed?

Rootlets from the dorsal and ventral surfaces of the spinal cord unite to form the dorsal and ventral roots of the spinal nerve. Upon exiting the vertebral canal, the dorsal and ventral roots join together to form the spinal nerve.

Does the dorsal primary ramus transmit motor or sensory information? What about the ventral primary ramus?

The dorsal and ventral primary rami convey both sensory and motor information. In contrast, dorsal and ventral nerve roots convey 1 type of fiber or the other.

What type of information does the dorsal root of a spinal nerve transmit, and in what direction?

Sensory ("afferent") information, toward the spinal cord

What type of information does the ventral root of a spinal nerve transmit, and in what direction?

Motor ("efferent") information, away from the spinal cord

Describe the location of the sensory dermatome associated with each of the following dorsal roots:

C6   Radial forearm and thumb ("six-shooter")

T4   Nipple

T10   Umbilicus

L1   Inguinal **L1**gament

S3   Genitoanal region

Which muscles are supplied by ventral roots from:

C3–C5?   The diaphragm. This explains why spinal cord injuries above this level are often fatal ("C3, C4, and C5 keep the diaphragm alive").

C8–T1?   Muscles of the hand

S1–S2?   Ankle plantar flexors. S1 nerve root compression from a herniated L5–S1 nucleus pulposus ("slipped disc") leads to characteristic weakness of plantarflexion.

S3–S5?   Muscles of the bladder, anal sphincter, and genitals

SPINAL TRACTS

What is a spinal tract?   A bundle of axons (i.e., white matter) connecting parts of the CNS

Through which spinal tracts are tactile, vibratory, and proprioceptive (joint position) sense transmitted from the spinal cord to the brain?   The fasciculus gracilis and fasciculus cuneatus (ascending tracts of the **dorsal column,** named after its location in the dorsal spinal cord)

What type of sensory information does the spinothalamic tract transmit from the spinal cord to the thalamus?

Pain and temperature

Where is the spinothalamic tract located?

In the anterolateral aspect of the spinal cord

Which spinal tract conveys motor information from the cerebral cortex to the spinal cord?

The corticospinal tract (aka the "pyramidal tract"), located in the lateral aspect of the spinal cord

VASCULATURE OF THE SPINAL CORD

What arteries constitute the arterial blood supply of the spinal cord?

1. The unpaired anterior spinal artery
2. 2 posterior spinal arteries
3. The radicular ("root") branches of the vertebral, cervical, posterior intercostal, and lumbar arteries

The anterior spinal artery is a branch of which artery?

The 2 vertebral arteries (They combine short branches to become the single midline anterior spinal artery.)

Which radicular artery provides the main blood supply to the inferior two thirds of the spinal cord?

The arteria radicularis magna, or major anterior radicular artery, clinically known as the "artery of Adamkiewicz." This artery may be damaged with aortic surgery, leading to associated damage to the spinal cord.

MENINGES OF THE SPINAL CORD

Name the meningeal layers that surround the spinal cord.

The meninges that surround the spinal cord are continuations of those that surround the brain (i.e., the pia mater, arachnoid mater, and dura mater).

What is the filum terminale?

The filum terminale, an extension of the pia mater that attaches to the coccyx, represents the most caudal extension of the spinal cord tissue.

POWER REVIEW

CNS

What is the name of a collection of nerve cell bodies located:	
Within the CNS?	Nucleus
Outside of the CNS?	Ganglion

BRAIN

Which structure joins the cerebral hemispheres?	The corpus callosum
What structures are separated from one another by the lateral sulcus (i.e., the sylvian fissure)?	The lateral sulcus separates the temporal lobe from the frontal and parietal lobes.
Which portion of the brain is concerned with:	
Voluntary motor activity?	The precentral gyrus
Sensory data?	The postcentral gyrus
Where is the visual cortex located?	In the occipital lobe
What is the primary function of the cerebellum?	Coordination of movement
What are the 3 major parts of the brainstem?	The midbrain (mesencephalon), pons, and medulla oblongata
What are the 3 layers of meninges, from superficial to deep?	Dura mater, arachnoid mater, pia mater
Name the 4 dural folds that subdivide the brain.	Falx cerebri, tentorium cerebelli, falx cerebelli, diaphragma sellae
What are the 2 midline dural folds?	The falx cerebri and the falx cerebelli

Describe the falx cerebri.   The falx cerebri is the dural fold located in the longitudinal fissure between the 2 cerebral hemispheres.

Describe the tentorium.   The tentorium cerebelli is the horizontal dural fold that supports the occipital lobes and covers the cerebellum.

What is the term "supratentorial" used to refer to?   Structures located above the tentorium cerebelli

Name the branches of the internal carotid artery within the neck.   There are none.

Each internal carotid artery ends by dividing into which 2 arteries?   The anterior cerebral artery and the middle cerebral artery

Through which intracranial artery does the majority of blood from the internal carotid artery flow?   The middle cerebral artery

The vertebral arteries join to form which structure?   The basilar artery

How does the basilar artery end?   By dividing into the left and right posterior cerebral arteries

How does blood travel from the great cerebral vein (of Galen) back to the heart?   After entering the straight sinus, the blood passes to the transverse sinus and then to the sigmoid sinus, which drains into the internal jugular vein. The internal jugular vein drains into the brachiocephalic vein, which in turn empties into the superior vena cava.

Which spaces communicate through the:

Interventricular foramen (of Monroe)?   The 2 lateral ventricles and the third ventricle

Cerebral aqueduct (of Sylvius)?   The third ventricle and the fourth ventricle

Foramen of Magendie (median aperture)?	The fourth ventricle and the cerebromedullary cistern (i.e., the cisterna magna)
Foramina of Luschka (lateral apertures)?	The fourth ventricle and the cerebromedullary cistern (i.e., the cisterna magna)
What structure separates the lateral ventricles?	The septum pellucidum
What is the source of CSF?	Choroid plexus within cerebral ventricles

SPINAL CORD

State where each of the following spinal nerves exits the vertebral canal:	
Spinal nerve C1	Above vertebra C1 (i.e., the atlas)
Spinal nerve C8	Below vertebra C7
How many spinal nerves and cranial nerves leave the CNS?	31 spinal nerves and 12 cranial nerves ("Eat 31 flavors in 12 months.")
Name the sensory (dermatomal) level:	
Nipples?	T4
Umbilicus?	T10
Inguinal ligament?	L1
Which artery that supplies the upper portion of the spinal cord is paired?	The posterior spinal artery (a branch of the vertebral artery). The anterior spinal artery is unpaired.
Which artery constitutes the principal arterial supply of the inferior two thirds of the spinal cord?	The major anterior radicular artery (of Adamkiewicz)
How many vertebrae comprise each level of the vertebral column?	7 cervical, 12 thoracic, 5 lumbar, 5 sacral (fused), 4 coccygeal (fused)

Where does the adult spinal cord end?

At vertebral level L2

What are the functions of the:

    Spinothalamic (anterolateral) tracts?

Convey pain/temperature sensation

    Dorsal columns?

Convey light touch, vibration, proprioception (position sense)

    Corticospinal tracts?

Convey motor fibers

---

4   The Cranial Nerves

What quality makes a nerve a cranial nerve (CN)?

To be a cranial nerve, a nerve must pass through a foramen in the skull.

How many pairs of cranial nerves are there?

12

Give the names of the 12 cranial nerves:

    I

Olfactory nerve

    II

Optic nerve

    III

Oculomotor nerve

    IV

Trochlear nerve

    V

Trigeminal nerve

    VI

Abducens nerve

    VII

Facial nerve

    VIII

Vestibulocochlear nerve

    IX

Glossopharyngeal nerve

X	Vagus nerve
XI	Spinal accessory nerve
XII	Hypoglossal nerve

For each cranial nerve, state whether it carries sensory fibers, motor fibers, or both:

I (olfactory nerve)	Sensory
II (optic nerve)	Sensory
III (oculomotor nerve)	Motor
IV (trochlear nerve)	Motor
V (trigeminal nerve)	Both
VI (abducens nerve)	Motor
VII (facial nerve)	Both
VIII (vestibulocochlear nerve)	Sensory
IX (glossopharyngeal nerve)	Both
X (vagus nerve)	Both
XI (spinal accessory nerve)	Motor
XII (hypoglossal nerve)	Motor
	Some Say Marry Money, But My Brother Says Big Brains Matter More

Which cranial nerves contain *special sensory* fibers?	1. CN I (olfactory nerve)
	2. CN II (optic nerve)
	3. CN VII (facial nerve)
	4. CN VIII (vestibulocochlear nerve)
	5. CN IX (glossopharyngeal nerve)
	6. CN X (vagus nerve)

Which cranial nerves carry *parasympathetic* **fibers?**	1. CN III (oculomotor nerve) 2. CN VII (facial nerve) 3. CN IX (glossopharyngeal nerve) 4. CN X (vagus nerve)
Are these parasympathetic fibers pre- or postganglionic?	Preganglionic
Which cranial nerves carry preganglionic sympathetic fibers?	None
Which cranial nerves carry all types of fibers (sensory, motor, parasympathetic)?	1. CN VII (facial nerve) 2. CN IX (glossopharyngeal nerve) 3. CN X (vagus nerve)
How are cranial nerve fibers *most accurately* **character-ized by functional type?**	1. Fibers may be **general** or **special.** 2. Fibers may be **visceral** or **somatic.** 3. Fibers may be **afferent** (toward brain, sensory) or **efferent** (away from brain, motor). For example, fibers that medi-ate the sensation of smell are classified as special visceral afferent (SVA).
For each cranial nerve or group of cranial nerves, state where the nerve or group of nerves emerges from the brain:	
I	Anterior forebrain
II	Diencephalon
III and IV	Midbrain
V	Pons
VI, VII, and VIII	Junction of the pons and the medulla
IX, X, XI (cranial root), and XII	Medulla
Where does the spinal root of CN XI (the spinal acces-sory nerve) originate?	From the upper segments of the spinal cord (C1–C6)

Which 3 cranial nerves do not originate on the brainstem?	1. CN I (olfactory nerve) 2. CN II (optic nerve) 3. CN XI (spinal accessory nerve)
Cranial nerves exit which aspect of the brainstem?	The ventral (anterior) surface
What is the 1 exception?	CN IV (the trochlear nerve) exits on the dorsal surface.

For each cranial nerve, state how it enters or exits the cranial cavity:

I (olfactory nerve)	Cribriform plate
II (optic nerve)	Optic canal
III (oculomotor nerve)	Superior orbital fissure
IV (trochlear nerve)	Superior orbital fissure
V (trigeminal nerve)	V_1—superior orbital fissure V_2—foramen rotundum V_3—foramen ovale
VI (abducens)	Superior orbital fissure
VII (facial nerve)	Internal acoustic meatus
VIII (vestibulocochlear nerve)	Internal acoustic meatus
IX (glossopharyngeal nerve)	Jugular foramen
X (vagus nerve)	Jugular foramen
XI (spinal accessory nerve)	Jugular foramen (Note that spinal root *enters* cranial cavity via foramen magnum.)
XII (hypoglossal nerve)	Hypoglossal canal
Which 2 cranial nerves are not peripheral nervous tissue?	CN I (the olfactory nerve) and CN II (the optic nerve)

Identify each of the labeled structures on the following figure of the base of the brain:

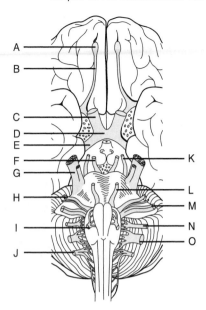

A = CN I (the olfactory nerve)
B = Olfactory tract
C = CN II (the optic nerve)
D = Optic chiasm
E = Optic tract
F = CN III (the oculomotor nerve)
G = CN V (the trigeminal nerve)
H = CN VII (the facial nerve)
I = CN XII (the hypoglossal nerve)
J = CN XI (the spinal accessory nerve)
K = CN IV (the trochlear nerve)
L = CN VI (the abducens nerve)
M = CN VIII (the vestibulocochlear nerve)
N = CN IX (the glossopharyngeal nerve)
O = CN X (the vagus nerve)

What is the only cranial nerve that extends into the abdomen?

CN X (vagus nerve)

Identify each of the labeled cranial nerves on the following figure of the base of the skull:

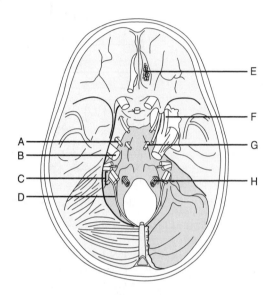

A = CN IV (the trochlear nerve)
B = CN V (the trigeminal nerve)
C = CN IX (the glossopharyngeal nerve), CN X (the vagus nerve), and CN XI (the spinal accessory nerve)
D = CN XI (the spinal accessory nerve)
E = CN I (the olfactory nerve)
F = CN V_2 (the maxillary division of the trigeminal nerve)
G = CN VI (the abducens nerve)
H = CN XII (the hypoglossal nerve)

Which cranial nerves are particularly vulnerable to injury during carotid endarterectomy (removal of atherosclerotic plaque from inner wall of internal carotid artery)?

1. CN X (vagus nerve)—located posteriorly within the carotid sheath
2. CN XII (hypoglossal nerve)—crosses internal carotid artery, just above carotid bifurcation
3. Mandibular branch of CN VII (facial nerve)—located along lower border of mandible

4. CN IX (glossopharyngeal nerve)—crosses internal carotid artery, high in the neck; injury uncommon

Which cranial nerves are most commonly affected with cavernous sinus thrombosis?

CN III (oculomotor nerve), CN IV (trochlear nerve), CN V (trigeminal nerve), and CN VI (abducens nerve). All these nerves lie in close proximity to this sinus.

CN I (OLFACTORY NERVE)

What type of fibers are carried by CN I?

Special sensory fibers (SVA)

What is the function of CN I?

Provides for the sense of smell

Describe the origin and course of the olfactory nerve from the nose

Approximately 20 neurosensory cells unite in the superior nasal cavity to form the small nerve bundles that comprise the olfactory nerve. These bundles pass through foramina in the cribriform plate of the ethmoid bone to enter the olfactory bulbs in the anterior cranial fossa. From the olfactory bulbs, impulses are conveyed to cortical centers by the olfactory tracts.

What is the most common cause of anosmia (i.e., an inability to smell)?

Chronic rhinitis. Other causes include fracture of the cribriform plate and tumors or abscesses of the frontal lobe that compress the olfactory bulb.

CN II (OPTIC NERVE)

What type of fibers are carried by CN II?

Special sensory fibers (SSA)

What is the function of CN II?

Vision

How does CN II enter the orbit?

Through the optic canal

Which structures accompany the optic nerve in the optic canal?

The ophthalmic artery and central vein of the retina

What term is used to describe the part of CN II in the visual pathway between the retina and the junction of the 2 optic nerves?

The optic *nerve*

What term is used to describe the part of the visual pathway where the 2 optic nerves from either side unite?

The optic *chiasm*

What term is used to describe the part of the visual pathway between the optic chiasm and the thalamus?

The optic *tract*

What part of the visual pathway connects the thalamus and the visual cortex?

The optic *radiations*

Trace visual input from the *lateral* retina to the thalamus.

Visual impulses from the *lateral* retina are conveyed in the *lateral* aspect of the optic nerve. At the optic chiasm, these impulses are conveyed to the *ipsilateral* optic tract en route to the thalamus (i.e., visual impulses from the lateral retina *do not cross* to the opposite side at the chiasm).

Which 3 sheaths enclose the optic nerve?

The optic nerve is enclosed by the 3 layers of cerebral meninges (pia, arachnoid, and dura).

The majority of the optic tract terminates in which structure?

The lateral geniculate body of the thalamus

The optic radiations connect which 2 structures?

The lateral geniculate body and the visual cortex (occipital)

Lesions at the optic nerve lead to what visual deficit?

Unilateral blindness or decreased acuity

Trace visual input from the
***medial* retina to the thalamus.**

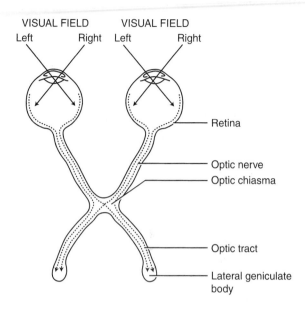

Impulses from the *medial* retina are conveyed in the *medial* aspect of the optic nerve. At the optic chiasm, these fibers "decussate" (cross) to the *contralateral* (opposite side) optic tract, where they run medially en route to the thalamus.

Lesions at the optic chiasm lead to what visual deficit?

Loss of vision in the temporal (lateral) visual field of both eyes, due to impingement on crossing fibers from both optic nerves. This lesion is termed "bitemporal hemianopsia" and is a classic presentation for a patient with a pituitary tumor.

Lesions beyond the optic chiasm lead to what visual deficit?

Lesions involving the optic tract, optic radiations, and cerebral cortex produce loss of vision within similar areas of the visual field in both eyes ("homonymous hemianopsia"). These lesions can be caused by stroke, tumor, vascular malformations, demyelinating lesions, and abscesses.

CN III (OCULOMOTOR NERVE)

Which types of fibers are carried by CN III?

Somatic motor (GSE) and visceral motor (GVE) fibers

What are the functions of CN III?

Somatic motor fibers: Innervate the medial, inferior, and superior recti muscles, the inferior oblique muscle, and the levator palpebrae superioris muscle

Visceral motor fibers: Provide visceral motor (parasympathetic) innervation to the sphincter pupillae and ciliary muscles

CN III travels between which 2 arteries on the base of the brain?

The posterior cerebral and superior cerebellar arteries

CN III passes through the lateral wall of which sinus?

The cavernous sinus

CN III enters the orbit through which opening?

The superior orbital fissure

List 3 other structures that also pass through the superior orbital fissure.

1. CN IV (the trochlear nerve)
2. CN V₁ (the ophthalmic division of the trigeminal nerve)
3. CN VI (the abducens nerve) (also ophthalmic veins)

 What are some common causes of CN III dysfunction?

1. Herniation of the brain through the foramen magnum (uncal herniation), as a result of increased intracranial pressure
2. Aneurysms
3. Fractures or inflammatory processes involving the cavernous sinus
4. Strokes
5. Multiple sclerosis

By what mechanism might brain swelling from trauma or a tumor lead to CN III dysfunction?

These conditions may increase intracranial pressure, leading to uncal herniation, with resultant compression of CN III. This leads to the classic clinical finding of a dilated, unresponsive ("blown") pupil ipsilateral to the side of brain herniation (loss of parasympathetic innervation to sphincter pupillae).

What structures are innervated by the 2 *divisions* of the oculomotor nerve?

Superior division—superior rectus and levator palpebrae superioris muscles

Inferior division—inferior and medial rectus, inferior oblique, ciliary muscle, and sphincter pupillae muscles

CN IV (TROCHLEAR NERVE)

Which types of fibers are carried by CN IV?

Somatic motor fibers (GSE)

What is the function of CN IV?

Motor innervation of the superior oblique muscle, which moves the eyeball inferiorly and laterally (in isolation) and inferiorly (in conjunction with the inferior rectus muscle)

What is the trochlea?

A fibrocartilaginous structure attached to the frontal bone that serves as a pulley for the tendon of the superior oblique muscle

CN IV exits the skull through what opening?

The superior orbital fissure. Remember: CN III (the oculomotor nerve), CN V_1 (the ophthalmic division of the trigeminal nerve), and CN VI (the abducens nerve) also pass through this fissure!

What is unique about CN IV, in terms of its relationship to the base of the brain?

It is the only cranial nerve to exit the brainstem on the dorsal surface.

Lesions of CN IV lead to what clinical deficit?

Failure of eye to depress fully, most obvious during adduction, with slight upward deviation of eye at rest

CN V (TRIGEMINAL NERVE)

CN V has 3 main divisions. Identify the territory of each on the following figure:

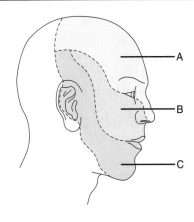

A = CN V_1 (the ophthalmic division)
B = CN V_2 (the maxillary division)
C = CN V_3 (the mandibular division)

Which types of fibers are carried by the trigeminal nerve?

Somatic sensory (GSA) and branchial motor (SVE) fibers

Which division of CN V carries all of the motor fibers?

CN V_3

CN V_3 provides motor innervation to which muscles?

1. The muscles of mastication (i.e., the medial and lateral pterygoids, the masseter muscle, and the temporalis muscle)
2. The tensor tympani muscle
3. The tensor veli palatini muscle
4. The mylohyoid muscle
5. The anterior belly of the digastric muscle

Where does each division of CN exit the middle cranial fossa?	**CN V₁:** The superior orbital fissure **CN V₂:** The foramen rotundum **CN V₃:** The foramen ovale

CN V₁: The superior orbital fissure
CN V₂: The foramen rotundum
CN V₃: The foramen ovale

What are the 3 major branches of CN V₁?

1. Frontal nerve (sensation to part of scalp, forehead, upper eyelid)
2. Nasociliary nerve (to eyeball, ethmoid sinuses)
3. Lacrimal nerve (to lacrimal gland) (all sensory branches)

What are the major branches of CN V₂?

1. Meningeal branches (to dura)
2. Superior alveolar nerves (to maxillary sinus, upper teeth)
3. Facial branches (to nose, upper lip, lower eyelid)
4. Nasal branches (to nasal mucosa) (all sensory branches)

What are the major branches of CN V₃?

1. Meningeal branches (to dura)
2. Buccal nerve (to cheek and buccal mucosa)
3. Auriculotemporal nerve (to temporomandibular joint [TMJ], external acoustic meatus, skin of temple, parotid gland)
4. Lingual nerve (to submandibular gland, sensation to anterior two thirds of tongue, floor of mouth)
5. Inferior alveolar nerves (to lower teeth, skin of chin/lower lip)

What part of the face is *not* innervated by CN V?

The angle of the mandible (innervated by spinal nerves C2 and C3)

CN VI (ABDUCENS NERVE)

CN VI carries which type of fibers?

Somatic motor (GSE)

What is the function of CN VI?

The abducens nerve provides motor innervation to the lateral rectus muscle (which abducts the eye).

CN VI exits the skull through which opening?	The superior orbital fissure (Remember: CN III, CN IV, and CN V_1 also pass through this fissure.)
Lesions of CN VI lead to what clinical deficit?	Failure of eye to abduct

CN VII (FACIAL NERVE)

Which types of fibers are carried by the facial nerve?	Somatic sensory (GSA), special visceral sensory (taste, SVA), visceral motor (parasympathetic, GVE), and branchial motor (SVE) fibers
Which muscles receive motor innervation from CN VII?	1. The muscles of facial expression 2. The stapedius muscle 3. The stylohyoid muscle 4. The posterior belly of the digastric muscle (Note the symmetry with CN V_3, which innervates the muscles of *mastication*, the tensor tympani muscle, the tensor veli palatini muscle, the mylohyoid muscle, and the *anterior belly* of the digastric muscle.)
What are the sensory functions of CN VII?	**Visceral (special) sensory functions:** Taste (anterior two thirds of tongue) and sensory innervation of the soft and hard palates **Somatic sensory functions:** Sensory innervation of the auricle and a small portion of skin behind the ear
Which structures receive parasympathetic innervation from CN VII?	1. The submandibular and sublingual salivary glands (via the chorda tympani branch) 2. The lacrimal glands (via the greater petrosal branch) 3. The secretory glands of the nasal and palatine mucosa (via the greater petrosal branch)

What are the only 2 sets of glands in the head that CN VII does *not* innervate?

1. The parotid gland (Note that although CN VII passes through the parotid gland, it *does not innervate* it.)
2. The integumentary glands (i.e., those of the scalp)

Trace the path of CN VII from the brainstem to its exit from the skull.

CN VII emerges from the caudal pons, courses through the internal acoustic meatus with CN VIII (the vestibulocochlear nerve), continues through the facial canal in the petrous portion of the temporal bone, and passes through the stylomastoid foramen to exit the skull.

Within which structure does the facial nerve divide?

The parotid gland

What are the 5 major terminal branches of CN VII?

Temporal, zygomatic, buccal, mandibular, and cervical

What is the function of the greater petrosal nerve?

This branch of CN VII carries parasympathetic fibers to the nasal and palatine mucosa, and to the lacrimal gland

Describe the course of the greater petrosal nerve.

After originating from the geniculate ganglion, the greater petrosal nerve exits the petrous part of the temporal bone through the hiatus of the greater petrosal canal and travels over the foramen lacerum. It then joins the deep petrosal nerve to form the nerve of the pterygoid canal, which passes through the pterygoid canal and then contributes to the pterygopalatine ganglion.

Where does the chorda tympani branch away from CN VII?

In the descending part of the facial canal

What is the function of this branch?

1. Provides taste to the anterior two thirds of the tongue
2. Provides parasympathetic innervation to the submandibular and sublingual salivary glands

The chorda tympani exits the skull via which opening?

The petrotympanic fissure

Which nerve does the chorda tympani join after exiting the petrotympanic fissure?

The lingual nerve, a branch of CN V_3

What is the other branch from CN VII within the facial canal (besides the chorda tympani and greater petrosal nerves)?

The nerve to the stapedius muscle

Lesions of the facial nerve proximal to the internal acoustic meatus lead to which deficits? Why?

1. Unilateral facial paralysis (loss of facial muscle innervation, aka Bell's palsy)
2. Unilateral loss of taste in the anterior two thirds of the tongue (loss of chorda tympani taste fibers)
3. Decreased salivation (loss of innervation to the submandibular and sublingual glands)
4. Unilateral hyperacusis (loss of innervation to the stapedius muscle of the inner ear, which normally dampens sounds)

Lesions of the facial nerve distal to the stylomastoid foramen lead to what deficits?

Unilateral facial paralysis only. Taste, salivation, and dampening of sound are spared because the branches that innervate the tongue, salivary glands, and stapedius muscle arise proximal to this point.

How can one clinically differentiate between an upper motor neuron lesion of CN VII (e.g., stroke at level above ganglion) and a lower motor neuron lesion (e.g., Bell's palsy)?

The upper facial muscles have some innervation from both right and left sides of the motor cortex. Therefore, upper motor neuron lesions are associated with sparing of some ipsilateral upper facial muscle function. In contrast, lower motor neuron lesions produce a complete facial paralysis (upper and lower).

CN VIII (VESTIBULOCOCHLEAR NERVE)

CN VIII carries which types of fibers?	Special sensory fibers (SSA, like CN I, CN II, CN VII, CN IX, and CN X)
What is the function of the vestibulocochlear nerve?	Hearing (cochlear portion) and balance (vestibular portion)
Where does CN VIII exit the brainstem?	Branches representing the vestibular and cochlear portions emerge separately at a groove between the pons and medulla.
Which organs are innervated by the vestibular branch?	The utricle, saccule, and semicircular canals (i.e., the 3 organs responsible for maintaining equilibrium)
In addition to balance, what else does the vestibular branch of CN VIII mediate?	Coordination of head and eye movements
Which organ is innervated by the cochlear branch?	The cochlea
Where does CN VIII exit the posterior cranial fossa?	Through the internal acoustic meatus (in the petrous portion of the temporal bone)
Which other structures travel with CN VIII through the internal acoustic meatus?	CN VII (i.e., the facial nerve) and the labyrinthine (internal acoustic) artery, a branch from the basilar artery

CN IX (GLOSSOPHARYNGEAL NERVE)

Which types of fibers are carried by CN IX?	Somatic sensory (GSA), visceral sensory (GVA), special sensory (taste, SVA), visceral motor (parasympathetic, GVE), and branchial motor (SVE)
What are the motor functions of CN IX?	**Branchial motor fibers:** Innervate the stylopharyngeus muscle, which elevates the pharynx during swallowing and speech **Visceral motor (parasympathetic) fibers:** Supply the otic ganglion, which provides secretomotor fibers to the parotid gland

What are the sensory functions of CN IX?

Somatic sensory fibers: Provide sensation to the upper pharynx, tonsils, posterior third of the tongue, skin of the external ear, and the internal portion of the tympanic membrane

Special sensory fibers: Provide the posterior third of the tongue with taste sensation

Visceral sensory fibers: Carry afferent input from the carotid body and sinus

Which structure receives parasympathetic innervation from CN IX?

The parotid gland (Remember: The parotid gland is traversed by CN VII but innervated by CN IX.)

Where does CN IX exit the cranium?

At the jugular foramen (located at the suture line between the inferior edges of the temporal and occipital bones)

Which other cranial nerves exit the skull at the jugular foramen?

CN X (the vagus nerve) and CN XI (the spinal accessory nerve)

What are the 6 major branches of CN IX?

1. Tympanic branch (to middle ear)
2. Carotid branch (to carotid body and sinus)
3. Pharyngeal branch (to pharyngeal mucosa)
4. Muscular branch (to stylopharyngeus)
5. Tonsillar branches
6. Lingual branch (to posterior tongue)

 What is the key to finding CN IX on dissection?

Locate the stylopharyngeus muscle. CN IX emerges posteriorly and runs around the lateral border of the muscle.

CN X (VAGUS NERVE)

Which types of fibers does CN X carry?

Branchial motor (SVE), visceral motor (GSE, parasympathetic), somatic sensory (GSA), and visceral sensory (GVA) fibers

What are the motor functions of CN X?

Branchial motor fibers: Innervate the striated muscles of the pharynx, palate, and larynx (except for the stylopharyngeus and tensor veli palatini muscles), and the palatoglossus muscle of the tongue

Visceral motor fibers: Innervate the smooth muscle of the thoracic and abdominal viscera

What are the sensory functions of CN X?

Somatic sensory fibers: Innervate the skin on the back of the ear, the external acoustic meatus, the external tympanic membrane, and the pharynx

Visceral sensory fibers: Innervate the larynx, trachea, and esophagus; the thoracic and abdominal viscera; the stretch receptors in the aortic arch; and the chemoreceptors in the aortic bodies

Which structures receive parasympathetic innervation via CN X?

1. The cardiac plexus (Parasympathetic innervation slows the heart and constricts coronary arteries.)
2. The pulmonary plexus (Parasympathetic innervation constricts the bronchial tree.)
3. The abdominal branches (Parasympathetic innervation provides for gastrointestinal motility as far as the left colic flexure and stimulates some gastrointestinal secretions.)

Describe the origin of CN X.

CN X arises as 8–10 rootlets from the medulla.

How does CN X exit the skull?

Via jugular foramen (with CN IX and CN XI)

List the 7 major branches of CN X.

1. Meningeal branch (provides sensation to the dura mater in the posterior cranial fossa)
2. Auricular branch (provides sensation to the back of the ear and communicates with the auricular branch of CN VII)
3. Superior laryngeal branch (The internal laryngeal branch provides sensation in the larynx superior to the vocal cords; the external laryngeal branch provides motor innervation to the inferior constrictor and cricothyroid muscles.)

4. Recurrent laryngeal branch (provides sensation to larynx inferior to vocal cords and innervates all of the intrinsic muscles of the larynx except the cricothyroid muscle)
5. Nerve to the carotid body and sinus
6. Motor branch to the pharyngeal plexus (supplies all striated muscles of pharynx and soft palate except stylopharyngeus and tensor veli palatini)
7. Parasympathetic branches to the thoracic and abdominal viscera

Where does CN X travel in the neck?

Within the carotid sheath, posterolateral to the carotid artery and posteromedial to the internal jugular vein

What structure does the recurrent laryngeal nerve loop around on the:
    Right side?

The subclavian artery

    Left side?

The ligamentum arteriosum (embryologic remnant of the ductus arteriosus, which connects the pulmonary artery with the aortic arch)

The vagus joins with which structure before leaving the thorax?

The contralateral vagus (forming the esophageal plexus and vagal trunks)

Which vagus nerve (right or left) forms the anterior vagal trunk?

The left
(Remember **LARP**: **L**eft **A**nterior, **R**ight **P**osterior)

How can the variation in the paths of the left and right vagus nerves be explained?

By recalling the embryologic development of the region. As the foregut rotates (clockwise from above), the vagus nerve (which is adherent to the esophagus) rotates with it.

List the 6 structures supplied by the anterior vagal trunk.

1. The anterior aspect of the stomach
2. Lesser omentum
3. Liver
4. Pylorus
5. Head of the pancreas
6. The first 2 parts of the duodenum

Which structure is supplied by the posterior vagal trunk?

The posterior aspect of the stomach

 What clinical deficit occurs with *unilateral* injury to the recurrent laryngeal branch of the vagus nerve (e.g., during thyroid or parathyroid surgery)?

Hoarseness

 What clinical deficit may occur with *bilateral* injury to the recurrent laryngeal branch of the vagus nerve?

Airway obstruction (inability to abduct either vocal cord), which may necessitate emergency tracheostomy (surgical placement of airway directly into trachea below vocal cords)

CN XI (SPINAL ACCESSORY NERVE)

CN XI carries which types of fibers?

Somatic motor (GSE) and branchial motor (SVE) fibers

What is the function of CN XI?

Spinal root: Provides motor innervation to the sternocleidomastoid and trapezius muscles

Cranial root: Provides branchial motor innervation to the larynx and pharynx via the pharyngeal and recurrent laryngeal branches of CN X

Describe the course of the spinal root of CN XI.

After originating from cervical segments C1–C6, the spinal root of CN XI travels superiorly into the cranium through the foramen magnum, joins the cranial root, and then exits the skull through the jugular foramen.

 The spinal accessory nerve may be injured during neck dissection to remove lymph nodes. What clinical deficit results from this injury?

Weakness when turning head to opposite side (function of sternocleidomastoid muscle) and drooping of shoulder (loss of trapezius function)

Where in the neck is the spinal accessory nerve particularly at risk for injury?

In the floor of the posterior triangle of the neck (i.e., between the trapezius and sternocleidomastoid)

CN XII (HYPOGLOSSAL NERVE)

CN XII carries which types of fibers?

Somatic motor fibers (GSE)

What is the function of CN XII?

CN XII provides motor innervation to all of the intrinsic and extrinsic muscles of the tongue, except for the palatoglossus muscle (which is supplied by CN X).

Where does CN XII originate?

CN XII emerges from the brainstem as 8–10 rootlets between the "olive" and the pyramid of the ventral medulla.

Where does CN XII exit the skull?

Through the hypoglossal canal (in the occipital bone)

Is CN XII medial or lateral to CN IX, CN X, and CN XI upon exiting the cranium?

Medial (Think of the tongue being in the middle.)

Which spinal nerve fibers travel with CN XII?

The descending branches of spinal nerve C1, which join with branches from spinal nerves C2 and C3 to form the *ansa cervicalis*. This structure supplies innervation to the sternohyoid, sternothyroid, and omohyoid muscles (all "strap" muscles of the anterior neck).

Describe the *physical examination* to assess function of each of the cranial nerves:
    I (olfactory nerve)

Test smell (not often actually performed, except by a neurologist! Most clinicians describe cranial nerve exam as testing cranial nerves II–XII.).

II (optic nerve)	Test visual acuity, visual fields, and pupillary light reflexes; perform fundoscopic examination.
III (oculomotor nerve)	Test extraocular movements, eyelid elevation, and pupillary constriction.
IV (trochlear nerve)	Test extraocular movements. Specifically, look for movement of both eyes out and down.
V (trigeminal nerve)	Test sensation on the face and palpate jaw as patient bites down.
VI (abducens nerve)	Test extraocular movements. Specifically, look for abduction of both eyes.
VII (facial nerve)	Test for upper and lower facial muscle function (e.g., smile, close eyes tightly).
VIII (vestibulocochlear nerve)	Test hearing (cochlear function). Vestibular function is not routinely tested.
IX (glossopharyngeal nerve)	Test gag reflex on both sides of pharynx. (Test this 1 last!)
X (vagus nerve)	Observe for midline palate/uvula as patient phonates (says "ah").
XI (spinal accessory nerve)	Have patient shrug his or her shoulders and turn his or her head to each side against resistance.
XII (hypoglossal nerve)	Have patient stick our his or her tongue. Look for symmetry.

POWER REVIEW

CRANIAL NERVES

For each cranial nerve, state the cranial foramen through which it passes:

    I (the olfactory nerve)   The cribriform plate

    II (the optic nerve)   The optic canal

    III (the oculomotor nerve)   The superior orbital fissure

    IV (the trochlear nerve)   The superior orbital fissure

    V_1 (the ophthalmic division of the trigeminal nerve)   The superior orbital fissure

    V_2 (the maxillary division of the trigeminal nerve)   The foramen rotundum

    V_3 (the mandibular division of the trigeminal nerve)   The foramen ovale

    VI (the abducens nerve)   The superior orbital fissure

    VII (the facial nerve)   The internal acoustic meatus and stylomastoid foramen

    VIII (the vestibulocochlear nerve)   The internal acoustic meatus

    IX (the glossopharyngeal nerve)   The jugular foramen

    X (the vagus nerve)   The jugular foramen

    XI (the spinal accessory nerve)   The jugular foramen (Note that spinal root enters skull through foramen magnum.)

    XII (the hypoglossal nerve)   The hypoglossal canal

Which 4 cranial nerves carry parasympathetic fibers?

1. CN III (the oculomotor nerve)
2. CN VII (the facial nerve)
3. CN IX (the glossopharyngeal nerve)
4. CN X (the vagus nerve)

Which 6 cranial nerves contain special sensory fibers?

1. CN I (olfactory nerve)
2. CN II (optic nerve)
3. CN VII (facial nerve)
4. CN VIII (vestibulocochlear nerve)
5. CN IX (glossopharyngeal nerve)
6. CN X (vagus nerve)

Lesions at the optic nerve lead to what visual deficit?

Unilateral blindness/decreased acuity

Lesions at the central optic chiasm lead to what deficit?

Loss of vision in temporal (lateral) visual field of both eyes ("bitemporal hemianopsia")

Lesions beyond the optic chiasm lead to what deficit?

Loss of vision in similar areas of visual field for both eyes ("homonymous hemianopsia")

Between which 2 arteries does CN III (the oculomotor nerve) emerge from the brain?

CN III emerges from the brain between the superior cerebellar and posterior cerebral arteries.

What is the function of CN IV (the trochlear nerve)?

Motor innervation to the superior oblique muscle

Lesions of CN IV lead to what clinical deficit?

Failure of eye to depress fully, especially noticeable during adduction, with slight upward deviation of eye at rest

What are the 3 main sensory divisions of CN V (the trigeminal nerve)?

1. **CN V$_1$:** The ophthalmic division
2. **CN V$_2$:** The maxillary division
3. **CN V$_3$:** The mandibular division

CN V provides motor innervation to which muscles?

1. The muscles of mastication
2. The tensor tympani muscle
3. The tensor veli palatini muscle
4. The mylohyoid muscle
5. The anterior belly of the digastric muscle

What is the function of CN VI (the abducens nerve)?

Motor innervation to the lateral rectus muscle, which abducts the eye

CN VII (the facial nerve) provides motor innervation for which structures?

1. The muscles of facial expression
2. The stapedius muscle
3. The stylohyoid muscle
4. The posterior belly of the digastric muscle

Which nerve provides for taste on the:

    Anterior two thirds of the tongue?

The chorda tympani, a branch of CN VII

    Posterior third of the tongue?

CN IX (the glossopharyngeal nerve)

Within what structure does the facial nerve divide?

The parotid gland

What are the 5 terminal branches of CN VII?

Temporal, zygomatic, buccal, mandibular, and cervical

What is the function of CN VIII (the vestibulocochlear nerve)?

Provides for hearing and equilibrium

What are the symptoms and signs of vestibular nerve dysfunction?

Dizziness, impaired balance, nystagmus, and nausea or vomiting

What muscle is innervated by CN IX (glossopharyngeal nerve)?

Stylopharyngeus

What nerve provides innervation to the parotid gland?

CN IX (glossopharyngeal nerve), *not CN VII!*

Where does CN X (the vagus nerve) travel in the neck?

Within the carotid sheath, posterolateral to the carotid artery and posteromedial to the internal jugular vein

The recurrent laryngeal nerve loops around which structure on the right? On the left?

The subclavian artery and the ligamentum arteriosum, respectively (Remember: Unilateral damage causes hoarseness; bilateral damage causes airway obstruction.)

Which vagus nerve (i.e., the right or the left) forms the anterior vagal trunk?

The left ·
(Remember **LARP**: **L**eft **A**nterior, **R**ight **P**osterior)

Which intrinsic muscle of the larynx is not supplied by the recurrent laryngeal nerve?

The cricothyroid muscle

Which nerve innervates the cricothyroid muscle?

The *superior laryngeal* branch of CN X (external branch)

What is the function of CN XI (the spinal accessory nerve)?

The spinal root supplies motor innervation to the sternocleidomastoid and trapezius muscles, and the cranial root supplies motor innervation to the larynx and pharynx.

The vagus nerve supplies all striated muscles of the pharynx and soft palate except which one?

Stylopharyngeus

What is the function of CN XII (the hypoglossal nerve)?

Moves the tongue (if CN XII is damaged, the tongue will deviate toward the side of the lesion)

CN XII provides motor innervation to all tongue muscles except for which one?

Palatoglossus muscle (supplied by CN X)

What nerve innervates the muscles of mastication?

CN V (trigeminal nerve, mandibular branch)

What nerve innervates the muscles of facial expression?

CN VII (facial nerve)

What is the only cranial nerve that extends into the abdomen?

CN X (vagus nerve)

What nerve provides secretomotor fibers to the lacrimal gland?

CN VII (facial nerve, via greater petrosal nerve)

What nerve provides sensory innervation to the paranasal sinuses?	CN V (trigeminal nerve)
What nerve innervates the middle ear cavity?	CN IX (glossopharyngeal)

5   The Neck

BONES OF THE NECK

What are the superior bony landmarks of the neck?	The inferior margin of the mandible anteriorly and the superior nuchal line of the occipital bone posteriorly
What are the inferior bony landmarks of the neck?	The superior borders of the clavicles and the manubrium (i.e., the superior bone of the sternum) anteriorly and the line connecting the 2 acromions posteriorly
What is the hyoid bone?	A small U-shaped bone located below the mandible (at the level of the C3 vertebral body) that serves as an attachment site for many muscles of the anterior neck
Which 2 muscles originate on the hyoid bone?	The hyoglossus (1 of the extrinsic tongue muscles) and the middle constrictor muscle of the pharynx

SUPERFICIAL MUSCLES OF THE NECK

What are the 2 superficial muscles of the neck?	1. The sternocleidomastoid muscle 2. The platysma Note that while the trapezius muscle extends into the neck, it is normally classified as a superficial *back* muscle.

🩺 **What is the most superficial muscle encountered when making an anterior incision in the neck (i.e., thyroidectomy, carotid endarterectomy)?**

The platysma

Platysma

What are the attachments of the platysma muscle?

This thin, flat muscle extends from the fascia of the pectoralis and deltoid muscles of the upper extremity to the mandible. (*Plat*- means "flat.")

What is the innervation of the platysma?

The platysma is innervated by the cervical branch of cranial nerve (CN) VII (the facial nerve).

What does the platysma do?

Assists in facial expressions, especially frowning

Sternocleidomastoid Muscle

Origin?

The mastoid process of the skull

Insertions?

As the sternocleidomastoid muscle runs anteromedially, it splits into 2 heads to insert on the sternum ("sterno") and the clavicle ("cleido").

Innervation?

The spinal branches of CN XI (the accessory spinal nerve) and spinal nerves C2 and C3

Actions?

Acting singly, the sternocleidomastoid muscle tilts the head to the ipsilateral side, bending the neck laterally and rotating the face so that it looks superiorly to the other side. Acting together, the sternocleidomastoid muscles flex the neck.

DEEP MUSCLES OF THE NECK

MUSCLES OF THE ANTERIOR TRIANGLE

What is the main function of the deep muscles of the anterior triangle?	To stabilize or move the hyoid bone and larynx, especially during phonation (speech)
Deep muscles in the anterior triangle can be divided into what 2 major groups?	1. Suprahyoid 2. Infrahyoid

Suprahyoid Muscles

Which 4 muscles comprise the suprahyoid group?	1. The mylohyoid 2. The geniohyoid 3. The stylohyoid 4. The digastric

Mylohyoid Muscle

Origin?	The mandible (mylohyoid line)
Insertion?	The hyoid bone (median raphe)
Innervation?	The mylohyoid nerve, a branch of CN V_3 (the mandibular branch of the trigeminal nerve)
Action?	This pair of thin, flat muscles form the muscular floor of the mouth and act to elevate the hyoid bone and tongue during speech (and also depress the mandible).

Geniohyoid Muscle

Origin?	The medial mandible (inferior mental spine)
Insertion?	The hyoid bone
Innervation?	Spinal nerve C1, via CN VII (the hypoglossal nerve)
Action?	Pulls the hyoid bone anteriorly

Stylohyoid Muscle

Origin?   The styloid process of the temporal bone

Insertion?   The hyoid bone

Innervation?   CN VII (the facial nerve)

Action?   Elevates the hyoid bone

Digastric Muscle

What is the origin of the   The digastric muscle has an anterior belly
digastric muscle's name?   and a posterior belly. "Digastric" means
                                     "2 bellies."

Describe the 2 origins of the   The anterior belly arises from the
digastric muscle.   mandible. The posterior belly arises
                                     from the temporal bone, just deep to
                                     the mastoid process.

Describe the insertion of the   The 2 bellies are connected by a tendon
digastric muscle.   held in a fascial sling attached to the hyoid
                                     bone.

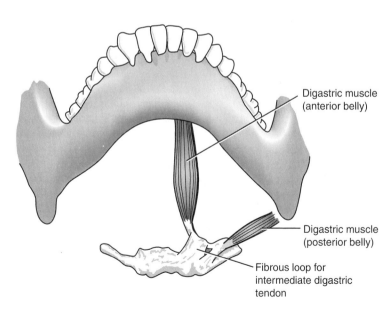

Digastric muscle
(anterior belly)

Digastric muscle
(posterior belly)

Fibrous loop for
intermediate digastric
tendon

Which 2 nerves innervate the digastric muscle?

Anterior belly: The mylohyoid nerve, a branch of the inferior alveolar nerve (from CN V$_3$)

Posterior belly: CN VII (the facial nerve)

Infrahyoid (Strap) Muscles

Which 4 muscles comprise the infrahyoid group?

1. The omohyoid
2. The sternohyoid
3. The sternothyroid
4. The thyrohyoid

What is the action of the strap muscles?

These muscles lower the hyoid and larynx during phonation and swallowing.

What nerve innervates all of the infrahyoid muscles but one?

The ansa cervicalis

Which infrahyoid muscle is not innervated by the ansa cervicalis?

The thyrohyoid muscle. This muscle is innervated by spinal nerve C1 via CN XII (the hypoglossal nerve).

Which strap muscle crosses perpendicular and just deep to the sternocleidomastoid muscle?

The omohyoid muscle

Describe the 2 heads of the omohyoid muscle.

As this muscle traverses from the scapula to the hyoid bone deep to sternocleidomastoid, it is tethered by a fascial sling connected to the clavicle, creating a superior and inferior belly.

Omohyoid Muscle

Origin?

Scapula (*omo-* means "shoulder" in Greek)

Insertion?

Hyoid bone

Sternohyoid Muscle

Origin?

Manubrium of the sternum

Insertion?

Hyoid bone

Sternothyroid Muscle

Origin?   Manubrium of the sternum

Insertion?   The thyroid cartilage

Thyrohyoid Muscle

Origin?   The thyroid cartilage

Insertion?   The hyoid bone

Muscles of the Posterior Triangle Floor

What separates the trapezius   Deep cervical (prevertebral) fascia
and sternocleidomastoid mus-
cles from the deep muscles of
the posterior triangle floor?

Which 4 deep muscles form   1. Splenius capitis muscle
the floor of the posterior   2. Levator scapulae muscle
triangle?   3. Posterior scalene muscle
                                      4. Middle scalene muscle

Which other muscle may   The anterior scalene muscle
contribute to the
inferomedial part of the
posterior triangle?

Where does the anterior   The first rib
scalene muscle insert?

Splenius Capitis Muscle

Origin?   The inferior half of the nuchal ligament
                                      and the spinous processes of vertebrae
                                      C1–C6

Insertion?   The mastoid process and the lateral
                                      superior nuchal line

Action?   **Unilateral contraction:** Flexes and
                                          rotates the head and neck ipsilaterally
                                      Bilateral contraction: Extends the head
                                          and neck

Identify the muscles of the neck on the following figure:

A = Digastric muscle, posterior belly
B = Stylohyoid muscle
C = Hyoglossus muscle
D = Mylohyoid muscle
E = Digastric muscle, anterior belly
F = Thyrohyoid muscle
G = Omohyoid muscle, superior belly
H = Sternothyroid muscle
I = Sternohyoid muscle
J = Sternocleidomastoid muscle
K = Anterior scalene muscle
L = Splenius capitis muscle
M = Levator scapulae muscle
N = Trapezius muscle
O = Posterior scalene muscle
P = Middle scalene muscle
Q = Omohyoid muscle, inferior belly

Levator Scapulae Muscle

Origin?

The transverse processes of vertebrae C1–C4

Insertion?

The superomedial border of the scapula

Innervation?	The dorsal scapular nerves and spinal nerves C3 and C4
Action?	Elevates and rotates the scapula (counterclockwise from the back)

Posterior Scalene Muscle

Origin?	The transverse processes of vertebrae C4–C6
Insertion?	The second rib
Innervation?	Spinal nerves C7 and C8
Action?	Flexes the neck laterally and elevates the second rib during inspiration

Middle Scalene Muscle

Origin?	The transverse processes of vertebrae C2–C7
Insertion?	The first rib
Innervation?	Spinal nerves C3–C8
Action?	Flexes the neck laterally and elevates the first rib during inspiration

FASCIAE OF THE NECK

SUPERFICIAL CERVICAL FASCIA

What is the superficial fascia of the neck called?	The investing fascia
What does the investing fascia invest?	The investing fascia is found on both sides of (i.e., it "invests") the sternocleidomastoid muscle. It runs inferior to the platysma and surrounds all of the deeper structures of the neck.

DEEP CERVICAL FASCIAE

**What are the 3 deep fasciae
of the neck?**

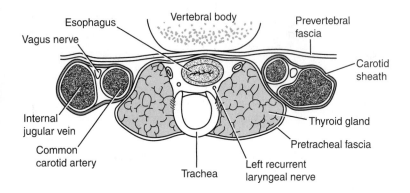

Vertebral body

Esophagus

Prevertebral
fascia

Vagus nerve

Carotid
sheath

Internal
jugular vein

Thyroid gland

Common
carotid artery

Pretracheal fascia

Trachea

Left recurrent
laryngeal nerve

1. Pretracheal fascia
2. Prevertebral fascia
3. Fascia of the carotid sheath

**What does the pretracheal
fascia enclose?**

Located deep to the infrahyoid muscles,
this fascia surrounds the trachea, thyroid
glands, and esophagus.

**What does the inferior
border of the pretracheal
fascia merge with?**

The fibrous pericardium of the
mediastinum

**What does the prevertebral
fascia cover?**

It envelops the vertebrae and the adjacent
deep cervical muscle masses, including
the floor of the posterior triangle.

**What is the retropharyngeal
space?**

The retropharyngeal space is a potential
space between the prevertebral fascia
and the pretracheal fascia.

TRIANGLES OF THE NECK

**Which muscle divides each
side of the neck into anterior
and posterior triangles?**

The sternocleidomastoid

Which structures delineate the anterior triangle of the neck?

Anterior boundary: Median line of the neck
Posterior boundary: The anterior border of the sternocleidomastoid muscle
Base: Inferior mandible
Apex: Jugular notch (i.e., the space between the clavicles and above the manubrium, aka the sternal notch)

Identify the labeled structures and triangles on the following illustration:

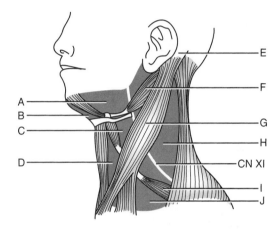

A = Digastric (submandibular triangle)
B = Submental triangle
C = Carotid triangle
D = Muscular triangle
E = Mastoid process
F = Digastric muscle (posterior belly)
G = Sternocleidomastoid muscle
H = Occipital triangle
I = Omohyoid muscle (posterior belly)
J = Supraclavicular triangle

Which 2 muscles divide the anterior triangle into 4 smaller triangles?

The digastric and superior belly of the omohyoid muscles

What are the 4 subdivisions of the anterior triangle?

1. The submental triangle
2. The digastric (submandibular) triangle
3. The carotid triangle
4. The muscular triangle

Which structures delineate the posterior triangle of the neck?

Anterior boundary: The posterior border of the sternocleidomastoid muscle
Posterior boundary: The anterior border of the trapezius muscle
Base: The middle third of the clavicle
Apex: Point where the sternocleidomastoid and trapezius muscles meet, on the occipital bone

What are the 2 subdivisions of the posterior triangle?

1. Occipital triangle
2. Supraclavicular triangle

What muscle divides the posterior triangle into 2 smaller triangles?

Inferior belly of the omohyoid

ANTERIOR TRIANGLE

Submental Triangle

What is unique about the submental triangle?

It is the only unpaired triangle within the neck.

What are its boundaries?

It is bounded by the hyoid bone inferiorly and the anterior bellies of the left and right digastric muscles laterally.

What forms the floor of the submental triangle?

The 2 mylohyoid muscles

Which structures of interest are found in the submental triangle?

The submental lymph nodes, which drain the tip of the tongue, the lower incisors, and the lower lip and chin

Digastric (Submandibular) Triangle

What bounds the digastric triangle?

The inferior margin of the mandible (superior) and the superior margins of both the anterior and posterior bellies of the digastric muscle (inferior)

What structure nearly fills the digastric triangle?

The submandibular gland

Which major nerve passes through the digastric triangle?

CN XII (the hypoglossal nerve), supplying motor innervation to the tongue muscles

Carotid Triangle

What are the boundaries of the carotid triangle?

The anterior border of the sternocleido-mastoid muscle (posterior), the superior belly of the omohyoid muscle (medial), and the posterior belly of the digastric muscle (superior)

Identify the following structures in the carotid and digastric triangles:

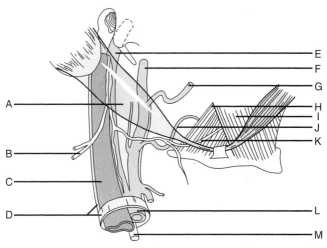

A = Digastric muscle, posterior belly
B = CN XI (spinal accessory nerve)
C = Internal jugular vein
D = Carotid sheath
E = Internal carotid artery
F = External carotid artery
G = Facial artery
H = Hyoglossus muscle
I = Mylohyoid muscle
J = Lingual artery
K = CN XII (hypoglossal nerve)
L = Common carotid artery
M = CN X (vagus nerve)

Which important structures pass through the carotid triangle?

The common carotid artery (and its bifurcation into internal and external branches), the internal jugular vein, and the vagus nerve (CN X). All these structures are found within the carotid sheath. (Occasionally, the superior part of the ansa cervicalis descends within the sheath as well.)

Muscular Triangle

What are the boundaries of the muscular triangle?

The anterior border of the sternocleidomastoid muscle posteriorly, the superior belly of omohyoid superiorly, and the median plane of the neck medially

What are the principal contents of the muscular triangle?

The infrahyoid muscles, the thyroid gland, and the parathyroid glands

VASCULATURE OF THE NECK

ARTERIES

What arteries enter the root of the neck from the thorax:
  On the left side?

1. The left common carotid artery
2. The left subclavian artery

  On the right side?

The brachiocephalic trunk (innominate artery)

What is the origin of all these arteries?

The arch of the aorta

Which 2 arteries does the brachiocephalic trunk divide into?

1. The right common carotid artery
2. The right subclavian artery

Clinicians refer to the brachiocephalic trunk by what additional name?

The innominate artery

What structure lies directly posterior to the proximal innominate artery?

The trachea. Tracheoinnominate fistula is a rare life-threatening complication of tracheotomy (surgical airway directly into trachea for patients with laryngeo-facial injuries or on prolonged mechanical ventilation). This condition presents as bright red blood from the tracheostomy tube.

Subclavian Artery

How does the subclavian artery pass out of the thorax and into the neck?

Through the **thoracic outlet:** Posterior to the sternoclavicular joint and over the first rib, between the anterior and middle scalenus muscles, and under the clavicle. In "thoracic outlet syndrome," the subclavian artery or vein may be compressed by scalenus muscle hypertrophy or a cervical (extra) rib.

Which structure divides the subclavian artery, and into what parts?

The subclavian artery is divided into 3 parts as it passes behind the anterior scalenus muscle. The first part is medial to the muscle, the second part is posterior to it, and the third part is lateral to it.

Which nerve loops under the right subclavian artery?

The right recurrent laryngeal nerve, a branch of CN X (the vagus nerve)

What are the 5 major branches off the subclavian artery?

1. The internal thoracic (internal mammary) artery
2. The vertebral artery
3. The thyrocervical trunk
4. The costocervical trunk
5. The dorsal scapular artery

All but which of these branches arise from the *first part* of the subclavian artery?

All branches arise from the first part of the subclavian artery, except for the costocervical trunk on the right and the dorsal scapular artery on the left, both of which usually arise from the second part.

Describe the course and fate of the vertebral arteries.

The vertebral arteries travel in the vertebral foramina of C6 and above, wind around the lateral masses of the atlas (C1), and enter the skull via foramen magnum. Vertebral arteries from either side then join posterior to the pons to form the basilar artery. (Recall that the vertebrobasilar system, along with the internal carotid system, provides blood to the brain.)

What are the 3 branches of the thyrocervical trunk?

1. The inferior thyroid artery
2. The transverse cervical artery
3. The suprascapular artery

What does the costocervical trunk divide into?

The superior intercostal and deep cervical arteries

What does the subclavian artery become as it leaves the neck and travels distally? When does this occur?

The subclavian artery becomes the axillary artery as it passes the lateral border of the first rib.

Common Carotid Artery

Describe the course of the common carotid artery.

Ascends from the root of the neck within the carotid sheath (extension of prevertebral fascia), posterior to the sternocleidomastoid and into the carotid triangle

What 2 other structures are found within the carotid sheath?

1. Internal jugular vein
2. Vagus nerve (CN X)

What 2 nerves in the carotid triangle are particularly at risk for injury during carotid endarterectomy (removal of atherosclerotic plaque from internal carotid artery to prevent stroke)?

1. Vagus nerve (CN X)
2. Hypoglossal nerve (CN XII)

How does the common carotid artery terminate?

By dividing into the internal and external carotid arteries

Where does the common carotid bifurcate?

Within the carotid triangle, at the superior border of the thyroid cartilage

What is the carotid body? What is its function?

A small ovoid mass of tissue located at the bifurcation of the common carotid artery. The carotid body contains *chemoreceptors* that detect changes in arterial O_2 and CO_2 levels (responds to low O_2 or high CO_2 by stimulating respiration centrally).

What is the carotid sinus?

A dilatation of the internal carotid artery just distal to the carotid bifurcation; contains *pressure* receptors

Which nerve innervates the carotid sinus?

CN IX (the glossopharyngeal nerve)

What are the branches of the internal carotid artery within the neck?

There are none! The first branch of the internal carotid artery is the ophthalmic artery, which arises intracranially.

What are the 2 terminal branches of the external carotid artery?

The maxillary artery and the superficial temporal artery

What are the 6 branches of the external carotid artery before this bifurcation occurs, in ascending order?

1. Superior thyroid artery
2. Ascending pharyngeal artery
3. Lingual artery
4. Facial artery
5. Occipital artery
6. Posterior auricular artery

Of these 6 branches, which are:

    Anterior?

1. The superior thyroid artery
2. The lingual artery
3. The facial artery
(Arguably the most important 3)

    Posterior?

1. The occipital artery
2. The posterior auricular artery

    Medial?

The ascending pharyngeal artery

Identify the arteries of the neck on the following figure:

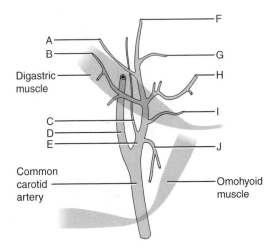

A = Posterior auricular artery
B = Occipital artery
C = Ascending pharyngeal artery
D = Internal carotid artery
E = External carotid artery
F = Superficial temporal artery
G = Maxillary artery
H = Facial artery
I = Lingual artery
J = Superior thyroid artery

VEINS

External Jugular Vein

Which veins join to form the external jugular vein?

The retromandibular and posterior auricular veins

Into what structure does the external jugular vein drain?

Subclavian vein (not internal jugular)

Internal Jugular Vein

What areas are drained by the internal jugular vein?

Structures in the cranium and the face

Where does the internal jugular vein originate?

A direct continuation of the sigmoid sinus, the internal jugular vein exits the skull through the jugular foramen.

Which fascia covers the internal jugular vein?

The carotid sheath (which also invests the carotid artery and CN X)

What is the relation between the 3 structures within the carotid sheath?

The vein is lateral, the artery is medial, and the nerve is posterior. When placing a central venous line into the internal jugular (IJ) vein, the target for the needle should be just lateral to the carotid pulse.

NERVES OF THE NECK

SYMPATHETIC TRUNK

From where do the sympathetic nerve fibers in the neck arise?

From the superior thoracic spinal nerves (Remember: Sympathetic system is "thoracolumbar.")

How do sympathetic fibers ascend in the neck?

Preganglionic fibers from the thoracic spinal nerves form 3 cervical sympathetic ganglia (inferior, middle, and superior), which make up the "sympathetic trunk."

Is the sympathetic trunk in the carotid sheath?

No, the sympathetic trunk travels alongside (outside and posterior to) the carotid sheath.

 Interruption of the sympathetic trunk in the neck causes which syndrome?

Horner's syndrome

What is Horner's syndrome?

Ptosis: Drooping of the eyelid occurs when innervation of the levator palpebrae is disrupted.

Miosis: The pupil is constricted owing to unopposed parasympathetic action of the sphincter pupillae muscle.

Anhydrosis: Lack of perspiration from loss of sympathetic function

**Which lung tumor
can cause Horner's
syndrome?**

A superior sulcus tumor (Pancoast tumor)

PARASYMPATHETIC NERVES

**What nerve is responsible
for the parasympathetic
innervation of the neck?**

The vagus (CN X) (Remember: Parasympathetic system is "craniosacral.")

**Where does the vagus nerve
travel in the neck?**

Posterior within the carotid sheath

**What branch of the vagus
ascends into the neck from
the thorax?**

The recurrent laryngeal nerve; runs in
the tracheoesophageal groove

**What structure does the
recurrent laryngeal nerve
loop around:**
    On the right?

Right subclavian artery

    On the left?

Ligamentum arteriosum (runs between
aortic arch and pulmonary artery)

CERVICAL PLEXUS

**The cervical plexus originates
from which nerves?**

The ventral primary rami of spinal nerves
C1–C4

**The *superficial* division of
the cervical plexus gives rise
to which 4 sensory branches?**

1. Lesser occipital
2. Great auricular
3. Transverse cervical
4. Supraclavicular

**The *deep* division of the
cervical plexus gives rise to
which branches?**

1. Ansa cervicalis
2. Phrenic nerve
3. Branches to longus capitis, longus
   colli, levator scapula, and scalenus
   medius (+/− branches to sternocleidomastoid and trapezius)

**What does the ansa
cervicalis innervate?**

The infrahyoid strap muscles (all except
thyrohyoid)

What is the course of this nerve?

This U-shaped nerve (ansa means "loop") originates from spinal nerves C1–C3 and runs just anterior to the carotid sheath. While this position makes the ansa cervicalis vulnerable to injury during surgery on the carotid artery, unilateral injury generally does not lead to any clinical deficit.

Which nerve innervates the diaphragm? What is its origin?

The phrenic nerve, which arises from spinal nerves C3–C5. (Because "C3, C4, and C5 keep the diaphragm alive, " spinal cord injuries above this level may be fatal.)

Does the phrenic nerve have a sensory component?

Yes. It is responsible for referred left shoulder pain with a subphrenic abscess or ruptured spleen.

What is the path of the phrenic nerve in the neck?

The phrenic nerve lies on the anterior surface of the anterior scalenus muscle to enter the thoracic inlet.

Between which muscles does the brachial plexus emerge from the deep part of the neck?

The anterior and medial scalenus muscles

VISCERA OF THE NECK

THYROID GLAND

What connects the left and right lobes of the thyroid gland?

The isthmus

What is the isthmus anterior to?

The second and third tracheal rings

Describe the dual origin of the blood supply to the thyroid.

The superior thyroid artery originates from the external carotid artery. The inferior thyroid artery is a branch of the thyrocervical trunk (a branch of the subclavian artery).

Which third thyroid artery exists in 10% of people?

In a small percentage of the population, the thyroid ima artery—an unpaired branch directly from the aorta, brachiocephalic trunk, or left common carotid artery—ascends anterior to the thyroid and supplies the isthmus. This third artery can be a source of serious bleeding in patients undergoing thyroid surgery.

Describe the venous drainage of the thyroid.

Superior and middle thyroid veins drain to the internal jugular vein; inferior thyroid vein to the brachiocephalic vein

Which important nerve runs with the inferior thyroid artery?

The recurrent laryngeal nerve. It is important to avoid this nerve during surgery. Unilateral injury leads to hoarseness and bilateral injury can lead to airway obstruction.

PARATHYROID GLANDS

How many parathyroid glands are normally present?

4 (i.e., left and right superior and inferior)

Describe the most common location of the parathyroid glands:
Superior?

Usually at the junction of the upper and middle third of the posterolateral thyroid gland (at cricothyroid junction)

Inferior?

More variable then superior parathyroid; most often found near inferior pole of thyroid but may be found in the thymus, mediastinum, or even within the carotid sheath

Where do the parathyroid arteries originate?

They are usually branches from the inferior thyroid arteries.

What is the most common indication for removal of 1 or more parathyroid glands (parathyroidectomy)?

Hyperparathyroidism, or excessive release of parathyroid hormone, leading to hypercalcemia. This condition most often results from a single enlarged parathyroid (adenoma) that is functioning independently. Resection of an adenoma generally results in complete cure.

Injury to the parathyroid glands results in what abnormality?

Hypocalcemia, which can lead to tetany (muscle spasm). The parathyroid glands are most often injured during thyroid surgery.

LARYNX

What is the colloquial name for the larynx?

The "voice box"

At what vertebral level is the larynx located?

At the level of vertebrae C3–C6

What 2 structures does the larynx connect?

The pharynx and the trachea

What are the 2 functions of the larynx?

1. Valve to prevent aspiration of swallowed food
2. Phonation

Laryngeal Cartilages

What tissue type comprises the laryngeal skeleton?

Cartilage

How many cartilages are there in the larynx?

9, 3 paired and 3 unpaired

Name the 3 unpaired cartilages.

1. The thyroid cartilage
2. The cricoid cartilage
3. The epiglottic cartilage

What is the anatomic term for the "Adam's apple?"

The laryngeal prominence (i.e., where the laminae of the thyroid cartilage meet in the median plane)

What is the derivation of the word thyroid?

Greek thyros ("shield")

Which cartilage is palpable inferior to the thyroid cartilage?

The cricoid cartilage

What is a cricothyroidotomy?

An emergency surgical airway for patients that cannot be intubated orotracheally or nasotracheally. It involves incising the cricothyroid membrane and passing an endotracheal tube directly into the trachea.

What is unique about the cricoid cartilage?

It is the only *complete* ring of cartilage in the larynx (and trachea).

Identify the labeled structures on the following figure of the larynx:

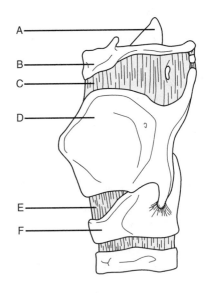

A = Epiglottic cartilage (i.e., epiglottis)
B = Body of the hyoid bone
C = Thyrohyoid membrane
D = Thyroid cartilage
E = Cricothyroid ligament
F = Cricoid cartilage

Describe the attachments of the epiglottis.

The leaf-shaped epiglottis is attached at its stalk to the thyroid cartilage, anteriorly to the hyoid bone, and laterally to the aryepiglottic folds.

Describe the function of the epiglottis.

It folds down over the entry of the larynx during swallowing to prevent aspiration.

Which ligaments anchor the cricoid cartilage?

Superiorly: The cricothyroid ligaments
Inferiorly: The cricotracheal ligaments

Which 2 structures pierce the thyrohyoid membrane on either side?

The internal laryngeal nerve and the superior laryngeal vessels

Name the 3 paired cartilages of the larynx.

1. Arytenoid cartilages
2. Corniculate cartilages
3. Cuneiform cartilages

Which of these pairs is most important in phonation, and how do they work?

The arytenoids. The vocal cords stretch from the thyroid cartilage to the pyramid-shaped arytenoid cartilages. These cartilages articulate with the cricoid cartilage posteriorly, allowing the position of (and the tension on) the vocal cords to change.

Where are the corniculate and cuneiform cartilages located?

The corniculate cartilages are located near the posterior apex of the arytenoid cartilages. The cuneiform cartilages are contained within the aryepiglottic folds.

Laryngeal Inlet

What are the aryepiglottic folds?

Folds of tissue containing cuneiform cartilage that form the lateral borders of the inlet to the larynx (the epiglottis forms the anterior border)

What are the piriform recesses?

Pear-shaped recesses ("*piriform*" = pearlike) lateral to the aryepiglottic folds

What are the valleculae?

2 small (peanut-sized) depressions formed by anterior attachments of the epiglottis; clinically significant because they are a frequent lodging place for food

Vocal Cords

What are the "true vocal cords"?

Vocal folds, consisting of the vocal ligaments and the conus elasticus

What is the conus elasticus?

An elastic membrane between the vocal ligaments and the cricoid cartilage. The vocal ligaments form the free edge of the conus elasticus.

Where do the vocal ligaments attach?

Between the vocal processes of the arytenoid cartilages and the posterior aspect of the thyroid cartilage

What is the sensory innervation of the larynx:
Above the vocal folds?

The internal branch of the superior laryngeal nerve, a branch of CN X (the vagus nerve)

Below the vocal folds?

The recurrent laryngeal branch of CN X

What are the "false vocal cords"?

Vestibular folds located superior to the "true" cords. These structures do not participate in sound production but appear similar to the true vocal cords.

Where are the vestibular folds located in relation to the vocal folds?

They are superior to the vocal folds.

Musculature of the Larynx

Which 2 groups of muscles comprise the laryngeal musculature?

1. Extrinsic muscles (move the entire larynx)
2. Intrinsic muscles (move the small cartilages and vocal cords)

Extrinsic muscles

What are the 3 muscles of the inlet of the larynx and what do they do?

1. **The transverse arytenoid muscle:** Joins the arytenoid cartilages (This muscle is the only unpaired muscle of the larynx.)
2. **The oblique arytenoid muscles:** Narrow the inlet
3. **The thyroepiglottic muscles:** Widen the inlet

Intrinsic Muscles

Which muscles are the principal:

    Adductors of the vocal cords?

The lateral cricoarytenoid muscles

    Abductors of the vocal cords?

The posterior cricoarytenoid muscles

    Tensors of the vocal cords?

The cricothyroid muscles

    Relaxers of the vocal cords?

The thyroarytenoid muscles

Which nerve innervates all but 1 of the intrinsic laryngeal muscles?

The recurrent laryngeal nerve

Which muscle is the exception, and which nerve innervates it?

The cricothyroid muscle is innervated by the external branch of the superior laryngeal nerve.

What is the path of the recurrent laryngeal nerves in the neck?

They travel down as part of CN X (the vagus nerve) within the carotid sheath, branch around the great vessels, and travel up the tracheoesophageal groove, finally piercing the cricothyroid muscle to enter the larynx.

POWER REVIEW

NECK

MUSCLES AND FASCIAE OF THE NECK

What are the 2 insertions of the sternocleidomastoid muscle?	The sternum and the clavicle
What is the most superficial muscle of the neck, and which nerve innervates it?	The platysma, innervated by CN VII (the facial nerve)
How are the anterior muscles of the neck classified?	As suprahyoid or infrahyoid
Which 4 muscles comprise the suprahyoid group?	1. Mylohyoid 2. Geniohyoid 3. Stylohyoid 4. Digastric
What are the origin, insertion, and innervation of the digastric muscle?	**Origin:** The temporal bone, deep to the mastoid process **Insertion:** Mandible **Innervation:** CN VII (posterior belly); CN V (anterior belly)
Which 4 muscles comprise the infrahyoid group?	1. Omohyoid 2. Thyrohyoid 3. Sternohyoid 4. Sternothyroid
What is the other name for this group of muscles?	The "strap" muscles
Which nerve innervates all of the strap muscles but one?	The ansa cervicalis
Which muscle is the exception, and which nerve innervates this muscle?	The thyrohyoid, which is innervated by spinal nerve C1 via CN XII (the hypoglossal nerve)

What 4 deep muscles form the floor of the posterior triangle?

1. Splenius capitis
2. Levator scapulae
3. Posterior scalenus
4. Middle scalenus

What travels in the carotid sheath?

The common carotid artery, the internal jugular vein, and CN X (the vagus nerve)

What travels just outside the carotid sheath?

The sympathetic trunk and the ansa cervicalis

Disruption of the sympathetic trunk causes what triad of symptoms?

Ptosis, miosis, and anhydrosis (i.e., Horner's syndrome)

What 2 structures pass between the anterior and middle scalenus muscles?

The brachial plexus and the subclavian artery

Vasculature of the Neck

What are the branches of the internal carotid artery in the neck?

There are none. The ophthalmic artery (in the cranium) is the first branch.

What are the 6 major branches of the external carotid artery, from proximal to distal?

1. Superior thyroid artery
2. Ascending pharyngeal artery
3. Lingual artery
4. Facial artery
5. Occipital artery
6. Posterior auricular artery

How does the external carotid artery terminate?

By dividing into the maxillary and superficial temporal arteries

What are the 3 major branches of the thyrocervical trunk?

1. Inferior thyroid artery
2. Transverse cervical artery
3. Suprascapular artery

INNERVATION OF THE NECK

What are the 4 major sensory branches of the cervical plexus?

1. Lesser occipital nerve
2. Great auricular nerve
3. Transverse cervical nerve
4. Supraclavicular nerve

What is the origin and course of the phrenic nerve?

After originating from spinal nerves C3–C5, the phrenic nerve travels on the anterior surface of the anterior scalenus muscle to enter the thoracic inlet.

VISCERA OF THE NECK

What is the arterial supply and venous drainage of the thyroid gland?

Arterial supply: Inferior and superior thyroid arteries (branches of the thyrocervical trunk and the external carotid arteries, respectively)

Venous drainage: The superior and middle thyroid veins drain to the internal jugular vein; the inferior thyroid vein drains to the brachiocephalic vein.

Where does the arterial supply of the parathyroid glands originate?

Usually from the inferior thyroid arteries

What is the anatomic term for the "true" vocal cords and where do they attach?

The *vocal folds* stretch between the vocal processes of the arytenoid cartilages and the posterior thyroid cartilage.

What is the anatomic term for the "false" vocal cords?

Vestibular folds

Which nerve innervates all but 1 of the intrinsic laryngeal muscles?

The recurrent laryngeal nerve; unilateral injury → hoarseness; bilateral injury → airway obstruction

What muscle is the exception, and which nerve innervates it?

The cricothyroid muscle, which is innervated by the superior laryngeal nerve

6

The Back

VERTEBRAE

What is the total number of adult vertebrae?	33
How many vertebrae are normally found in each region of the back?	**Cervical region:** 7 **Thoracic region:** 12 **Lumbar region:** 5 **Sacral region:** 5 (fused) **Coccygeal region:** 4 (fused)
What is the total number of spinal nerve roots on each side of the body?	31: 8 cervical, 12 thoracic, 5 lumbar, 5 sacral, and a single coccygeal nerve
Where do the cervical nerve roots exit the spine in relation to their respective vertebral bodies?	Superior to them (i.e., C1 nerve root runs superior to C1 vertebral body)
Where do the thoracic and lumbar nerve roots exit the spine in relation to their respective vertebral bodies?	Inferior to them (i.e., T7 nerve root runs inferior to C1 vertebral body)
Which 2 curves of the vertebral column are:	
Concave anteriorly?	The thoracic and sacral curves
Concave posteriorly?	The cervical and lumbar curves
Which 2 curves are known as the primary curvatures?	The thoracic and sacral curves
Why?	The thoracic and sacral curves develop in the fetal period and are present at birth.
Which 2 curves are known as the secondary curvatures?	The cervical and lumbar curves

When does the cervical curvature form?

When the infant holds its head erect (at 3–4 months)

When does the lumbar curvature form?

When the infant begins to walk (at the end of the first year)

 What is scoliosis?

Abnormal lateral curvature of the spine; present in 0.5% of population, more in females

 What is kyphosis?

Exaggerated posterior curvature of the spine, most often thoracic (leads to "humpback" appearance)

 What is lordosis?

Exaggerated anterior curvature of the spine, most often lumbar

Identify the labeled structures on the following views of a "typical" vertebra:

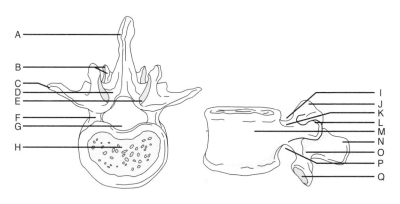

A = Spinous process
B = Inferior articular process and facet
C = Transverse process
D = Lamina
E = Superior articular facet
F = Pedicle
G = Vertebral foramen
H = Vertebral body

I = Superior vertebral notch
J = Superior articular process
K = Pedicle
L = Transverse process
M = Vertebral body
N = Spinous process
O = Lamina
P = Inferior vertebral notch
Q = Inferior articular facet

Which structures form the vertebral arch?

The pedicles (laterally) and the fused lamina (posteriorly). Laminectomy involves removal of this portion of a vertebra to gain access to the disk space and/or to decompress adjacent spinal nerve roots.

What is the function of the vertebral arch?

Protection of the spinal cord, nerve roots, and meninges

How many processes arise from the vertebral arch of a typical vertebra?

7 (2 transverse processes, 1 spinous process, and 4 articular processes)

What does the vertebral arch form, along with the vertebral body?

The vertebral foramen (the opening in each vertebra that permits passage of the spinal cord and meninges)

What is the vertebral canal?

The canal formed by the articulated vertebral foramina of successive vertebrae and the intervening intervertebral disks

What is the intervertebral foramen?

An opening between the pedicles of adjacent vertebrae

Which structures pass through the intervertebral foramen?

The exiting spinal nerve with its accompanying blood vessels

The spinous process arises from what portion of each vertebra?

The spinous process (posterior) is an extension of the 2 fused laminae.

Which cervical vertebra does not have a spinous process?

Vertebra C1

Describe the appearance of the spinous processes of vertebrae C2–C6.

Short and bifid

Why is vertebra C7 sometimes called the vertebra prominens?

It has the longest spinous process. Accordingly, this is the only cervical spinous process that is readily palpable.

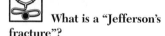 **What is a "clay-shoveler's" fracture?**

Fracture of the C7 spinous process

What are the alternative names for C1 and C2?

C1: Atlas (the atlas supports the skull, just as Atlas held up the globe)
C2: Axis

What are the unique characteristics of vertebra C1 (the atlas)?

Vertebra C1 lacks a body and a spinous process. The anterior and posterior arches connect the lateral masses that form its sides.

What is a "Jefferson's fracture"?

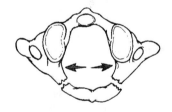

Fracture through C1 arches due to axial loading. This is a clinically unstable fracture.

What is the odontoid process (dens)?

The portion of vertebra C2 (the axis) that projects superiorly and articulates with the anterior arch of the atlas. An "odontoid" cervical spine X-ray is a specialized frontal X-ray view obtained with the patient's mouth widely open. This view is used to visualize the C1 lateral masses and the odontoid process.

How are fractures of the odontoid classified?

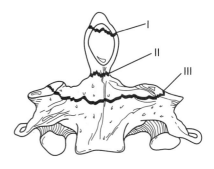

Type I: Through tip of dens (stable)
Type II: Through base of dens (unstable)
Type III: Through C2 body (generally stable)

What is the colloquial name for an odontoid fracture?

A hangman's fracture. This fracture through the pedicles of C2 usually results from hyperextension and is considered an "unstable" fracture.

What structure is unique to the cervical vertebrae?

The paired transverse foramina (located on either side of the vertebral body/lateral masses)

Which structures pass through the transverse foramina?

The vertebral artery, as well as the vertebral vein and autonomic nerves. The vertebral artery is vulnerable to injury with fractures of the transverse foramina.

Which cervical vertebra does not transmit vertebral arteries through its transverse foramina?

Vertebra C7

Which structures are unique to the thoracic vertebrae?	The costal facets (where the thoracic vertebrae articulate with the ribs)
What is unique about the lumbar vertebrae?	1. They have the largest bodies and pedicles. 2. A mamillary process is located on the posterior surface of each superior articular process. (Note: Vertebra T12 also has mamillary processes.)
At what vertebral level does the spinal cord usually terminate?	L2. However, spinal nerves L2–S5 continue caudally to exit by their corresponding vertebrae. This collection of nerve roots is referred to as the "cauda equina" (horse's tail).
What is the function of the sacrum?	The sacrum, which is formed by the 4–5 fused sacral vertebral bones, transmits body weight to the bony pelvis (via the sacroiliac joints).
What structures are transmitted by the sacral foramina (openings in anterior and posterior sacrum on each side)?	Sacral spinal nerves
What is the sacral promontory?	The anterior edge of vertebra S1 (forms the posterior boundary of the true pelvis)
What is the function of the coccyx?	Site of attachment for muscles and ligaments; does not contribute to weight-bearing

JOINTS AND LIGAMENTS

CRANIOVERTEBRAL JOINTS

What is the joint between the skull and vertebra C1 (the atlas) called?	The atlantooccipital joint (a synovial type joint)
What parts of C1 and the skull articulate at the atlantooccipital joint?	Occipital condyles of skull with superior articular facets of atlas

Which motion occurs at the atlantooccipital joint?	Flexion and extension of the head (nodding)
Which ligaments attach vertebra C1 to the skull?	The anterior and posterior atlantooccipital membranes
What is the joint between the atlas and the axis called?	The atlantoaxial joint (also a synovial joint)
What does the atlantoaxial joint consist of?	2 facet (plane) joints and 1 pivot joint between the dens and the anterior arch of the atlas
Which motion occurs at the atlantoaxial joint?	Rotation of the head from side to side. Remember the motions of the atlantooccipital and atlantoaxial joints by remembering "First yes (atlantooccipital), then no (atlantoaxial)."
Name the 3 major ligaments of the atlantooccipital joint.	1. Transverse ligament (C1) 2. Cruciform ligament 3. Alar ligaments
Where does the transverse ligament attach?	It runs between the tubercles on the lateral masses of vertebra C1, arching over the dens of vertebra C2
What is its purpose?	The transverse ligament holds the dens against the anterior arch of vertebra C1.
What are the points of insertion of the cruciform ligament?	**Horizontally:** The lateral masses of vertebra C1 **Superiorly:** The occipital bone **Inferiorly:** The body of vertebra C2
Where do the alar ligaments attach?	They run from the sides of the dens to the lateral margins of the foramen magnum
Which movement is prevented by the alar ligaments?	The alar ligaments check the rotation and side-to-side movement of the head.

JOINTS OF THE VERTEBRAL BODIES

Where are the intervertebral disks located?	Between the bodies of adjacent vertebrae

At what level is the most superior intervertebral disk found?

Between vertebrae C2 and C3 (There is no intervertebral disk between the atlas and the axis.)

At what level is the most inferior intervertebral disk found?

Between vertebrae L5 and S1

What is the external covering of the intervertebral disk called?

The anulus fibrosus

What is the internal matrix of the intervertebral disk called?

The nucleus pulposus. This gelatinous part of the intervertebral disk may become extruded (herniated nucleus pulposus, "slipped disk"). This characteristically causes back pain and symptoms related to local nerve root compression.

Herniation of the nucleus pulposus (HNP) most commonly occurs at which vertebral levels?

Lower cervical (C6–C7) and lower lumbar (L5–S1, L4–L5)

What are the signs of HNP at:
 C6–C7?

Triceps weakness (C7 root)

 L4–L5?

Weak foot extension/dorsiflexion (L5 root)

 L5–S1?

Weakness of plantarflexion (S1 root)

Which 2 ligaments play the most important role in stabilizing the vertebral bodies?

The anterior and posterior longitudinal ligaments

Where does the anterior longitudinal ligament run?

Along the anterior aspect of the vertebral bodies and intervertebral disks, from the occipital bone to the sacrum

Where does the posterior longitudinal ligament run?

Along the posterior aspect of the vertebral bodies, within the vertebral canal

INTERVERTEBRAL JOINTS

Which 2 structures form the facet joints?

The superior and inferior articular processes of adjacent vertebrae

What is the function of the facet joints?

The facet joints allow flexion, extension, and rotation of the spine. They also contribute to the spine's ability to bear weight and prevent anterior movement of the superior vertebra onto the inferior one.

Which ligament connects the lamina of adjacent vertebrae?

The ligamentum flavum

What does the ligamentum flavum do?

It contributes to the posterior boundaries of the intervertebral foramina and helps to straighten the vertebral column after flexion.

Identify the labeled ligaments and associated structures of the vertebral column on the following views:

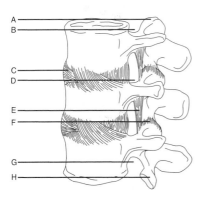

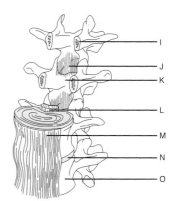

A = Superior articular process
B = Superior vertebral notch
C = Intervertebral disk
D = Intervertebral foramen
E = Ligamentum flavum
F = Articular capsule of the facet joint
G = Inferior vertebral notch
H = Inferior articular process
I = Pedicle
J = Ligamentum flavum
K = Lamina
L = Posterior longitudinal ligament
M = Anterior longitudinal ligament
N = Intervertebral disk
O = Vertebral body

When performing a lumbar puncture (spinal tap), which ligaments are pierced?

A lumbar puncture is usually performed at the level of vertebrae L3–L4 or L4–L5, in the midline between the iliac crests. After piercing the skin and superficial fascia, the needle passes through the supraspinous ligament, the interspinous ligament, and the ligamentum flavum before piercing the dura mater and the arachnoid mater to reach the cerebrospinal fluid (CSF).

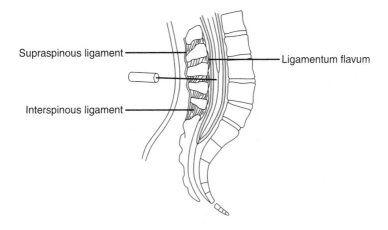

Supraspinous ligament

Ligamentum flavum

Interspinous ligament

Which ligaments connect the spinous processes?

The interspinous and supraspinous ligaments

What is the ligamentum nuchae?

A median fibrous septum between the posterior neck muscles. It is attached to the atlas and the cervical spinous processes and is the upward extension of the supraspinou ligament.

Which ligaments connect the transverse processes?

The intertransverse ligaments (These are most substantial in the lumbar region.)

MUSCLES

Identify the superficial muscles of the back and the related structures on the following figure:

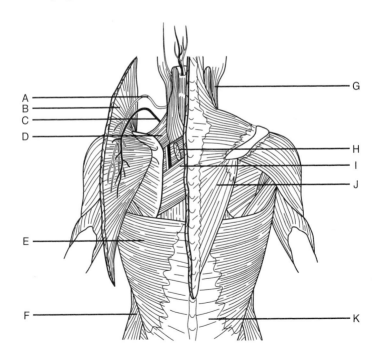

A = Accessory nerve
B = Trapezius muscle (reflected)
C = Transverse cervical artery
(superficial branch)
D = Levator scapulae muscle
E = Latissimus dorsi muscle
F = External abdominal oblique muscle
G = Sternocleidomastoid muscle
H = Rhomboid minor muscle (cut)
I = Rhomboid major muscle
J = Trapezius muscle
K = Thoracolumbar fascia

What are the 3 anatomic classifications used to categorize the muscles of the back?

Superficial, intermediate, or deep

What are the 2 functional classifications used to categorize the muscles of the back?

Extrinsic or intrinsic

How do these groups overlap?

Extrinsic = superficial and intermediate
Intrinsic = deep

Name the 5 superficial back muscles.

1. Trapezius
2. Latissimus dorsi
3. Levator scapulae
4. Rhomboid major
5. Rhomboid minor

Name the 3 intermediate back muscles.

1. Serratus posterior superior
2. Serratus posterior inferior
3. Levatores costarum

SUPERFICIAL BACK MUSCLES

Latissimus Dorsi Muscle

Origin?

The spinous processes of vertebrae T7–T12, the thoracolumbar fascia (to vertebrae L1–L5), ribs 9–12, the upper sacral vertebrae, and the iliac crest

Insertion?

The floor of the bicipital groove of the humerus

Innervation?

The thoracodorsal nerve (from the brachial plexus; receives branches from the C6, C7, and C8 ventral rami)

Action?

Adducts, extends, and rotates the humerus medially at the shoulder joint

Trapezius Muscle

Origin?

The external occipital protuberance, the superior nuchal line, the ligamentum nuchae, and the spinous processes of vertebrae C7–T12

Insertion?

The spine of the scapula, the acromion, and the lateral third of the clavicle

Innervation?

The spinal accessory nerve (cranial nerve [CN] XI) and branches of spinal nerves C3 and C4

Action?

Adducts, rotates, elevates, and depresses the scapula

Levator Scapulae Muscle

Origin?

The transverse processes of vertebrae C1–C4

Insertion?

The medial border of the scapula opposite the supraspinous fossa

Innervation?

The dorsal scapular nerve (from the brachial plexus; receives branches from the C5 ventral ramus)

Action?

Elevates the scapula

Rhomboid Minor Muscle

Origin?

The spinous processes of vertebra C7–T1

Insertion?

The root of the spine of the scapula

Innervation?

The dorsal scapular nerve

Action?	Adducts the scapula, fixes scapula to thoracic wall

Rhomboid Major Muscle

Origin?	The spinous processes of vertebrae T2–T5
Insertion?	The medial border of the scapula
Innervation?	The dorsal scapular nerve
Action?	Adducts the scapula; fixes scapula to thoracic wall

INTERMEDIATE BACK MUSCLES

Serratus Posterior Superior Muscle

Origin?	The ligamentum nuchae, the supraspinous ligament, and the spinous processes of vertebrae C7–T3
Insertion?	The upper border of ribs 2–5
Innervation?	Intercostal nerves T1–T4 (i.e., the T1–T4 ventral primary rami)
Action?	Elevates the ribs

Serratus Posterior Inferior Muscle

Origin?	The supraspinous ligament and the spinous processes of vertebrae T11–L3
Insertion?	The lower border of ribs 9–12
Innervation?	Intercostal nerves T9–T12 (i.e. the T9–T12 ventral primary rami)
Action?	Depresses the ribs

DEEP BACK MUSCLES

Name the 3 layers of deep muscles within the back.	1. Spinotransverse group (superficial) 2. Sacrospinalis group (intermediate) 3. Transversospinalis group (deep)

Which muscles comprise each group?

1. **Spinotransverse group:** Splenius capitis and splenius cervicis muscles
2. **Sacrospinalis group:** Erector spinae (formed by the iliocostalis, longissimus, and spinalis muscles)
3. **Transversospinalis group:** Semi-spinalis, multifidus, and rotatores muscles

Splenius Capitis Muscle

Origin?

The inferior half of the ligamentum nuchae and the spinous processes of vertebrae C7 and T1–T3

Insertions?

Temporal bone: At the mastoid process
Occipital bone: Along the lateral third of the superior nuchal line

Innervation?

The dorsal rami of the inferior cervical nerves

Actions?

Unilaterally: Ipsilateral lateral flexion and rotation of the head and neck
Bilaterally: Extension of the head and neck

Splenius Cervicis Muscle

Origin?

The spinous processes of vertebrae T3–T6

Insertion?

The transverse processes of vertebrae C1–C4

Innervation?

The dorsal rami of the inferior cervical nerves (the same as the splenius capitis muscle)

Actions?

Unilaterally: Ipsilateral lateral flexion and rotation of the head and neck
Bilaterally: Extension of the head and neck

Erector Spinae

Is the erector spinae palpable?

Yes. The 3 vertical columns (i.e., the iliocostalis, longissimus, and spinalis muscles) form the prominent bulge that is palpable along each side of the vertebral column.

How are the iliocostalis, longissimus, and spinalis muscles subdivided?

Into 3 parts each, according to the superior attachments (e.g., spinalis thoracis, spinalis cervicis, spinalis capitis)

Which fascial compartment encloses the erector spinae?

The erector spinae lies between the posterior and anterior layers of the thoracolumbar fascia.

What is the common origin of the erector spinae?

Most of the divisions of the columns attach through a broad tendon to:
1. The posterior part of the iliac crest
2. The posterior part of the sacrum
3. The sacroiliac ligaments
4. The sacral and lumbar spinous processes

What is the insertion for the:
    Iliocostalis muscle?

The ribs and cervical transverse processes

    Longissimus muscle?

The ribs, transverse processes, and mastoid process

    Spinalis muscle?

The spinous processes (Note that the spinalis also arises from the spinous processes.)

What is the action of the erector spinae:
    Unilaterally?

Lateral flexion of the head or vertebral column

    Bilaterally?

1. Extension of the vertebral column and the head
2. Control of movement during flexion

Semispinalis Muscle

What are the 3 divisions of the semispinalis muscle?
The semispinalis thoracis, the semispinalis cervicis, and the semispinalis capitis

What is the origin of the:

Semispinalis thoracis?
The thoracic vertebrae

Semispinalis cervicis?
The cervical vertebrae

Semispinalis capitis?
The occipital bone between the inferior nuchal lines

What is the insertion of the:

Semispinalis thoracis?
The spinous processes of vertebrae C6–T4

Semispinalis cervicis?
The spinous processes of vertebrae C2–C5

Semispinalis capitis?
The planum nuchale (occipital bone)

Which nerves innervate the semispinalis muscle?
The dorsal rami of the spinal nerves

What is the bilateral action of the semispinalis muscle?
Extension of the head and upper vertebral column

What is the unilateral action of the semispinalis muscle?
Contralateral rotation of the head

Multifidus Muscle

In which region is the multifidus muscle most prominent?
The lumbar region

Describe the origin and insertion of the multifidus muscle.
The multifidus muscle runs superomedially from the vertebral arches to the spinous processes, covering the laminae and spanning 3–4 vertebrae.

Innervation?
The dorsal rami of the spinal nerves

Actions?

Unilateral action: Ipsilateral flexion and contralateral rotation of the vertebral column

Bilateral action: Extension and stabilization of the spine

Rotatores Muscles

Describe the origin and insertion of the rotatores.

The rotatores arise from the transverse process of 1 vertebra and insert on the spinous process of the next (i.e., superior) vertebra.

Innervation?

The dorsal rami of the spinal nerves

Action?

Contralateral rotation and stabilization of the vertebral column

SUBOCCIPITAL REGION

What are the 4 major muscles of the suboccipital region?

1. Rectus capitis posterior major
2. Rectus capitis posterior minor
3. Obliquus capitis superior
4. Obliquus capitis inferior

What is the rectus capitis posterior major muscle's:
    Origin?

The spinous process of vertebra C2 (the axis)

    Insertion?

The lateral portion of the inferior nuchal line

What is the rectus capitis posterior minor muscle's:
    Origin?

The posterior tubercle of vertebra C2 (the axis)

    Insertion?

The medial part of the inferior nuchal line

What is the obliquus capitis superior muscle's:
    Origin?

The transverse process of vertebra C1 (the atlas)

Insertion?

The occipital bone, above the inferior nuchal line

What is the obliquus capitis inferior muscle's:
    Origin?

The spinous process of vertebra C2 (the axis)

    Insertion?

The transverse process of vertebra C1 (the atlas)

Which nerve innervates all of the suboccipital muscles?

The suboccipital nerve

Where does the suboccipital nerve originate?

The suboccipital nerve originates from the dorsal ramus of vertebra C1 (the atlas) and emerges between the vertebral artery (above) and the posterior arch of the atlas (below).

What actions do the suboccipital muscles perform as a unit?

1. Extension of the head (all 4 muscles)
2. Rotation of the head (all muscles except for the obliquus capitis superior)
3. Flexion of the head laterally (all muscles, except for the obliquus capitis inferior)

POWER REVIEW

BACK

What is the total number of adult vertebrae?	33
What is the first easily palpated vertebra?	C7 (the long spinous process is palpable at the base of the neck)
Which feature is unique to the cervical vertebrae?	The paired transverse foramina
Do all of the transverse foramina transmit a vertebral artery?	No. Only small accessory vertebral veins pass through the transverse foramina of vertebra C7.
What is the odontoid process (dens)?	The part of vertebra C2 (the axis) that projects superiorly from the vertebral body and articulates with vertebra C1 (the atlas)
Which action occurs at the:	
Atlantooccipital joint?	Flexion and extension of the head (nodding)
Atlantoaxial joint?	Lateral movement of the head (i.e., from side to side)
What does the cruciform ligament connect?	The vertical part connects vertebra C2 (the axis) to the foramen magnum. The horizontal portion spans vertebra C2 across the dens.
Where is the most superior intervertebral disk? The most inferior?	**Most superior:** Between vertebrae C2 and C3 **Most inferior:** Between vertebrae L5 and S1
What is the name of the external covering of the intervertebral disk? The soft internal part?	The anulus fibrosis (fibrous cartilage) and the nucleus pulposus (elastic cartilage), respectively

At what vertebral level does the spinal cord usually terminate?

L2. spinal nerves L2–S5 continue caudally as "cauda equina."

What are the 5 superficial muscles of the back?

1. Trapezius
2. Latissimus dorsi
3. Rhomboid major
4. Rhomboid minor
5. Levator scapulae (often considered with the upper limb)

Which 2 muscles comprise the spinotransverse (superficial) group of the deep back muscles?

The splenius capitis and splenius cervicis muscles

What comprises the sacrospinalis (intermediate) group of the deep back muscles?

The erector spinae (3 parts)

Which 3 muscle groups comprise the deep group of the deep back muscles?

1. The semispinalis muscles (i.e., the semispinalis capitis, the semispinalis cervicis, and the semispinalis thoracis muscles)
2. The multifidus muscle
3. The rotatores muscles

7   The Upper Extremity

BONES

What are the 5 regions of the upper limb, and which bones are found in each region?

1. **Pectoral girdle:** Clavicle and scapula
2. **Arm (brachium):** Humerus
3. **Forearm (antebrachium):** Ulna and radius
4. **Wrist (carpus):** Carpal bones (8)
5. **Hand (manus):** Metacarpal bones (5) and phalanges (14)

Identify the labeled bones on the following figure of the upper extremity:

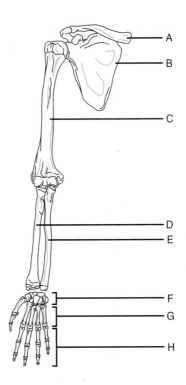

A = Clavicle
B = Scapula
C = Humerus
D = Radius
E = Ulna
F = Carpal bones
G = Metacarpal bones
H = Phalanges

PECTORAL GIRDLE

What is the function of the pectoral girdle?

The pectoral girdle connects the upper limb to the axial skeleton. (The axial skeleton consists of the skull, the vertebral column, the ribs and their cartilages, and the sternum.)

What are the 2 articulations of the clavicle?

The medial end of the clavicle articulates with the manubrium of the sternum at the sternoclavicular joint. The lateral end of the clavicle articulates with the acromion of the scapula at the acromioclavicular joint.

Describe the shape and surface anatomy of the clavicle.

1. Cylindrical and S-shaped with the medial two thirds of the clavicle being convex anteriorly. The large vessels and nerves that supply the upper limb pass posterior to the bone in this region.
2. Can be divided into thirds. Fractures are classified by which third is fractured.
3. The acromioclavicular joint can be palpated 2–3 cm medial to the acromion (i.e., the lateral extension of the spine of the scapula that forms the palpable "point" of the shoulder).

Where does the clavicle most frequently fracture?

Middle third: Narrowest portion and the only third without ligamentous attachments

What is the name of the anterior surface of the scapula?

The subscapular fossa

Describe the coracoid process.

This bony process arises from the superior border of the scapula; it is often described as having a "bird's beak" appearance.

Name the 6 structures that attach to the coracoid process.

3 muscles (pectoralis minor insertion, short head of biceps origin, coracobrachialis origin) and 3 ligaments (coracoacromial, coracoclavicular, coracohumeral)

Which bony landmark lies midway along the superior border of the scapula?

The suprascapular notch

Which ligament runs across the suprascapular notch?

The superior transverse scapular ligament

Which 2 structures traverse the superior transverse scapular ligament?

The suprascapular **a**rtery runs over the ligament, and the suprascapular **n**erve runs under the ligament. Think, "The **A**rmy goes over the bridge and the **N**avy goes under it."

Which structure divides the posterior surface of the scapula into 2 fossae?

The spine of the scapula divides the posterior surface of the scapula into the supraspinous fossa and the infraspinous fossa.

Which of these fossae is the largest?

The infraspinous fossa (i.e., the 1 below the spine of the scapula)

Which muscle originates from the:

Spine of the scapula?

The deltoid muscle

Supraspinous fossa?

The supraspinatus muscle

Infraspinous fossa?

The infraspinatus muscle

Subscapular fossa?

The subscapularis muscle

What are the articulations of the scapula?

The spine of the scapula continues laterally as the acromion, which articulates anteriorly with the clavicle to form the acromioclavicular joint. The lateral surface of the scapular body forms the glenoid fossa, which articulates with the head of the humerus to form the glenohumeral joint.

Dislocations at the glenohumeral joint occur most commonly in which direction?

Anterior, resulting in a subcoracoid location of the humeral head

What structure deepens the glenoid fossa?

The glenoid labrum, a fibrocartilaginous lip that extends over the glenoid fossa, thereby deepening it. Injury to this structure may predispose recurrent dislocation at the glenohumeral joint.

Which muscles originate just above and below the glenoid fossa?

The long head of the biceps brachii (supraglenoid tubercle) and the long head of the triceps brachii (infraglenoid tubercle), respectively

Identify the labeled structures on the following diagram of the scapula (posterior view):

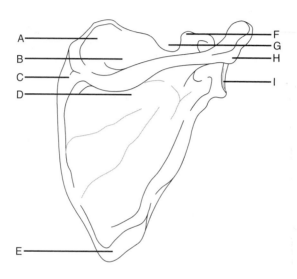

A = Superior angle
B = Supraspinous fossa
C = Root of the spine of the scapula
D = Infraspinous fossa
E = Inferior angle
F = Coracoid process
G = Suprascapular notch
H = Acromion
I = Glenoid cavity

Describe the surface anatomy of the scapula as it relates to the vertebral column.

1. The root of the spine of the scapula (i.e., the medial end) is opposite the spinous process of vertebra T3.
2. The superior angle of the scapula lies at the level of vertebra T2.
3. The inferior angle of the scapula lies at the level of vertebra T7.

What is the subacromial space and what is its clinical significance?

The subacromial space is located inferior to the acromion and superior to the humeral head. The muscles/tendons of the rotator cuff pass through this space and may be compressed at this location, particularly with overhead activity (shoulder "impingement").

ARM (BRACHIUM)

Identify the labeled structures on the following figure of the humerus (anterior view):

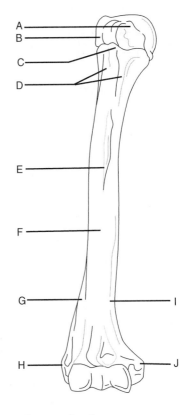

A = Lesser tubercle
B = Greater tubercle
C = Intertubercular groove
D = Surgical neck
E = Deltoid tuberosity
F = Humeral shaft (body)
G = Lateral supracondylar ridge
H = Lateral epicondyle
I = Medial supracondylar ridge
J = Medial epicondyle

Which 3 muscles insert on the greater tubercle of the humerus?

The supraspinatus, infraspinatus, and teres minor muscles (i.e., all the rotator cuff muscles except for the subscapularis muscle)

Which muscle inserts on the lesser tubercle of the humerus?

The subscapularis muscle (i.e., the fourth rotator cuff muscle)

Which 3 muscles insert on the intertubercular groove?

1. **Pectoralis major:** Lateral lip of the groove
2. **Teres major:** Medial lip of the groove
3. **Latissimus dorsi:** Floor of the groove. Remember: 3 "major" muscles insert on the intertubercular groove: The pectoralis major, the teres major, and the largest muscle of the back, the latissimus dorsi.

Where is the anatomic neck of the humerus?

Distal to the head of the humerus and proximal to the greater and lesser tubercles

Where is the surgical neck of the humerus?

Distal to the greater and lesser tubercles, where the humeral shaft begins

What is the most common site for proximal fractures of the humerus?

The surgical neck. This type of fracture commonly occurs in elderly osteoporotic women after a fall.

What are the borders of the *quadrangular* space?

Superior: Teres minor
Inferior: Teres major
Medial: Long head of triceps
Lateral: Humerus

What structures pass through the quadrangular space?

The axillary nerve and posterior circumflex humeral artery. Because of its proximity to the humerus at this level, the axillary nerve may be injured with fractures of the surgical neck.

What are the borders of the *triangular* space?

Superior: Teres minor
Inferior: Teres major
Lateral: Long head of triceps

What structure passes through the triangular space?

Branch of circumflex scapular artery

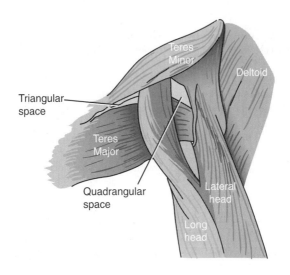

Triangular space

Teres Minor

Deltoid

Teres Major

Quadrangular space

Lateral head

Long head

Where is the radial (spiral) groove and why is it clinically important?

The radial groove runs obliquely along the posterior humerus and houses the radial nerve. Therefore, fractures affecting the shaft of the humerus may lead to radial nerve damage.

Name the 2 articular surfaces of the distal humerus.

1. The **capitulum** (i.e., the lateral articular surface) articulates with the head of the radius.
2. The **trochlea** (i.e., the medial articular surface) articulates with the trochlear notch of the ulna.

Where are the coronoid fossa and the olecranon fossa?

The coronoid fossa is superior to the trochlea on the distal end of the humerus anteriorly. The olecranon fossa lies in the same position posteriorly.

What structures do these fossae accommodate?

The coronoid process and the olecranon of the ulna, respectively

FOREARM (ANTEBRACHIUM)

Identify the labeled structures on the following figure of the bones of the forearm (anterior view):

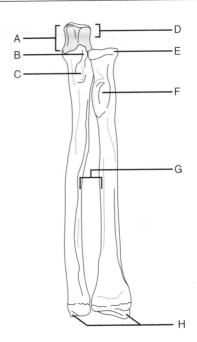

A = Trochlear notch
B = Coronoid process
C = Ulnar tuberosity
D = Olecranon
E = Head of the radius
F = Radial tuberosity
G = Interosseus borders
H = Styloid processes

Which muscle inserts on the styloid process of the radius?

The brachioradialis muscle

Which muscle inserts on the:
  Radial tuberosity?

The biceps brachii muscle

  Ulnar tuberosity?

The brachialis muscle

Describe the articulations of the:
  Proximal radius

The capitulum of the humerus and the radial notch of the ulna

Distal radius

The proximal row of carpal bones (except for the pisiform bone)

Where are the heads of the radius and ulna located?

The head of the radius is located at the proximal end of the bone, articulating with the capitulum of the humerus and the radial notch of the ulna. The head of the ulna is at the distal end of the bone, articulating with the articular disk of the radioulnar joint. To remember the location of the radial and ulnar heads, think **RPUD: R**adial **P**roximal, **U**lna **D**istal.

What attaches the lateral aspect of the shaft of the ulna and the medial aspect of the shaft of the radius?

The interosseous membrane

What other function does the interosseous membrane serve?

It is the attachment site for several of the deep forearm muscles.

WRIST (CARPUS)

Identify the labeled bones on the following figure of the wrist (posterior view):

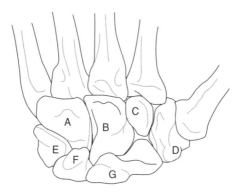

A = Hamate
B = Capitate
C = Trapezoid
D = Trapezium
E = Triquetrum
F = Lunate
G = Scaphoid

Note that the pisiform, which lies just anterior to the triquetrum, is not visible in a posterior view.

What is a mnemonic to remember the carpal bones?

From lateral to medial, proximal row followed by distal row:
Some **L**overs **T**ry **P**ositions **T**hat **T**hey **C**an't **H**andle
Scaphoid
Lunate
Triquetrum
Pisiform
Trapezium
Trapezoid
Capitate
Hamate
To remember that the trapezium comes before the trapezoid, think "trapezium with thumb."

What is the TFCC and what is its clinical significance?

The triangular fibrocartilage complex is a clinical term for the ligamentous and cartilaginous structures located on the ulnar side of the wrist between the distal ulna and proximal carpals. It acts to stabilize the radioulnar joint and absorb axial load in the wrist joint. Tears can result in ulnar-sided wrist pain.

Which of the 4 carpal bones in the proximal row does not articulate with the radius and articular disk?

The pisiform

Which of the 4 carpal bones in the proximal row articulates with the ulna?

None of the carpal bones articulates with the ulna.

Which carpal bone is the most commonly fractured?

The scaphoid. Fractures of this bone lead to tenderness over the "anatomic snuffbox," so named because this small compartment was a popular place to hold tobacco for snorting in the days when snuff was popular. The scaphoid's proximal pole has a tenuous blood supply and fractures may result in osteonecrosis.

HAND (MANUS)

Identify the labeled bones on the following figure of the wrist and hand (anterior view):

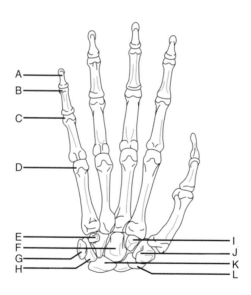

A = Distal phalanx
B = Head of the middle phalanx
C = Head of the proximal phalanx
D = Head of the fifth metacarpal bone
E = Hook of the hamate
F = Capitate
G = Pisiform
H = Triquetrum
I = Trapezoid
J = Tubercle of the trapezium
K = Lunate
L = Tubercle of the scaphoid

Name the 3 parts of a metacarpal bone and the location of each.

The base is proximal, the head is distal, and the shaft (body) is in the middle.

 What is a boxer's fracture?

Fracture of the fourth or fifth metacarpal, which often occurs after an object is punched with a closed fist

How many phalanges are in:

Each finger?

3 (proximal, middle, and distal)

The thumb?

2 (proximal, distal)

Which tendons of the fingers have sheaths and what is the clinical significance of infection in these tendon sheaths?

The flexor tendons are contained in sheaths while the extensor tendons are not. Poor vascularity makes infections within these sheaths (flexor tenosynovitis) a surgical urgency. While early cases may be treated with antibiotics, advanced infections need operative drainage.

PECTORAL GIRDLE AND SHOULDER

PECTORAL MUSCLES

Name the 4 pectoral muscles.

1. Pectoralis major muscle
2. Pectoralis minor muscle
3. Serratus anterior muscle
4. Subclavius muscle

Pectoralis Major Muscle

Origin?

Clavicular head: The anterior surface of the medial clavicle

Sternocostal head: The anterior surface of the sternum and the superior 6 costal cartilages

Insertion?

The lateral lip of the intertubercular groove of the humerus

Innervation?	The medial and lateral pectoral nerves (branches of the brachial plexus receiving fibers from the C8 and T1 ventral rami and the C5, C6, and C7 ventral rami, respectively)
Action?	Adducts and medially (internally) rotates the humerus at the glenohumeral (shoulder) joint
The superior border of the pectoralis major muscle contributes to which anatomic triangle?	The deltopectoral triangle (the other sides are formed by the deltoid and the clavicle)
What vascular structure lies in the deltopectoral triangle?	The cephalic vein (also the deltoid branch of the thoracoacromial artery)

Pectoralis Minor Muscle

Origin?	Ribs 2–5
Insertion?	The coracoid process of the scapula
Innervation?	The medial pectoral nerve
Action?	Stabilizes the scapula by drawing it anteriorly
The pectoralis minor muscle divides which important axillary structure into 3 parts?	The axillary artery
What is the clinical importance of the pectoralis minor during a mastectomy for breast cancer?	This muscle defines the 3 "levels" of lymph nodes within the axilla, the most common site for the spread of breast cancer. During a mastectomy for cancer the level I (lateral to pect minor) and II (deep to pect minor) nodes are removed, but the level III (medial to pect minor) nodes are usually left in place.

Serratus Anterior Muscle

Origin?

The external surfaces of ribs 1–8 laterally (The muscle is named for the saw-toothed appearance of its proximal attachments.)

Insertion?

The anteromedial border of the scapula

Innervation?

The long thoracic nerve (a branch of the brachial plexus receiving fibers from the C5, C6, and C7 ventral rami)

Action?

Holds the scapula against the thoracic wall

What happens when the long thoracic nerve is injured?

Loss of serratus anterior function, characteristically resulting in a phenomenon known as "winging of the scapula"

In addition to the long thoracic nerve, what other nerve in the axillary region is vulnerable to injury during lymph node dissection for breast cancer?

The thoracodorsal nerve, which innervates the latissimus dorsi muscle

What is the blood supply of the serratus anterior muscle?

The lateral thoracic artery (branch of part II of axillary artery); also receives some blood supply from the superior thoracic artery (branch of part I of axillary artery)

Subclavius Muscle

Origin?

Rib 1 and its costal cartilage

Insertion?

The inferior surface of the clavicle, at its middle third

Innervation?

The nerve to subclavius (a branch of the brachial plexus receiving fibers from the C5 and C6 ventral rami)

Action?

Stabilizes the clavicle by depressing it

SCAPULAR MUSCLES

Which 7 muscles pass from the scapula to the humerus and act on the shoulder joint?	1. Deltoid muscle 2. Teres major muscle 3. Supraspinatus muscle 4. Infraspinatus muscle 5. Teres minor muscle 6. Subscapularis muscle 7. Coracobrachialis
Which of these muscles comprise the "rotator cuff" muscles?	**SITS** **S**upraspinatus **I**nfraspinatus **T**eres minor **S**ubscapularis
Why are these 4 muscles known as the "rotator cuff" muscles?	Along with their corresponding tendons, these 4 muscles surround the glenohumeral joint, forming a musculotendinous "cuff" that protects and stabilizes the joint by holding the head of the humerus in the glenoid fossa.

Deltoid Muscle

Origin?	The lateral third of the clavicle, the acromion of the scapula, and the spine of the scapula
Insertion?	The deltoid tuberosity of the humerus
Innervation?	The axillary nerve (a terminal branch of the brachial plexus receiving fibers from the C5 and C6 ventral rami)
Action?	**Anterior part:** Flexion and medial (internal) rotation of the humerus at the glenohumeral joint **Middle part:** Abduction of the humerus at the glenohumeral joint **Posterior part:** Extension and lateral (external) rotation of the humerus at the glenohumeral joint

Teres Major Muscle

Origin?	The dorsal surface of the inferior angle of the scapula
Insertion?	The medial lip of the intertubercular groove of the humerus
Innervation?	The lower subscapular nerve (a branch of the brachial plexus receiving fibers from the C6 and C7 ventral rami)
Action?	Adduction and medial rotation of the humerus at the glenohumeral joint
What surface landmark is formed by the teres major muscle and the tendon of the latissimus dorsi muscle?	The posterior axillary fold

Supraspinatus Muscle

Origin?	The supraspinous fossa of the scapula
Insertion?	The superior facet of the greater tubercle of the humerus
Innervation?	The suprascapular nerve (a branch of the brachial plexus receiving fibers from the C4, C5, and C6 ventral rami)
Action?	Initiates and holds early abduction of the humerus at the glenohumeral joint; acts in concert with the other rotator cuff muscles to hold the head of the humerus in the glenoid fossa

Infraspinatus Muscle

Origin?	The infraspinous fossa of the scapula
Insertion?	The middle facet on the greater tubercle of the humerus
Innervation?	The suprascapular nerve

Action?   Lateral (external) rotation of the humerus at the glenohumeral joint; also assists in holding the head of the humerus in the glenoid fossa

Teres Minor Muscle

Origin?   The superior part of the lateral border of the scapula

Insertion?   The inferior facet on the greater tubercle of the humerus

Innervation?   The axillary nerve

Action?   Lateral (external) rotation of the humerus at the glenohumeral joint; also assists in holding the head of the humerus in the glenoid fossa (Note that the action of the teres minor muscle is identical to that of the infraspinatus muscle.)

Subscapularis Muscle

Origin?   The subscapular fossa on the anterior surface of the scapula

Insertion?   The lesser tubercle of the humerus

Innervation?   The upper and lower subscapular nerves (branches of the brachial plexus receiving fibers from the C5, C6, and C7 ventral rami)

Action?   Medial (internal) rotation and adduction of the humerus at the glenohumeral joint; also assists in holding the head of the humerus in the glenoid fossa

FASCIAE

The pectoral fascia is continuous with which structure inferiorly?   The fascia of the abdominal wall

Laterally, the pectoral fascia becomes what?   The axillary fascia

Which 2 muscles are enveloped by the clavipectoral fascia?

The subclavius and pectoralis minor muscles

What is the portion of the clavipectoral fascia between the first rib and the coracoid process of the scapula called?

The costocoracoid membrane

Which artery, vein, and nerve pierce the costocoracoid membrane?

The thoracoacromial artery (a branch of the axillary artery), the cephalic vein, and the lateral pectoral nerve

JOINTS AND LIGAMENTS OF THE PECTORAL GIRDLE

What are the 3 joints of the pectoral girdle?

1. Sternoclavicular joint
2. Acromioclavicular joint
3. Glenohumeral (shoulder) joint

Describe the sternoclavicular joint.

The sternoclavicular joint is a synovial joint. The synovial membrane lines the articular capsule between the sternal end of the clavicle and the manubrium of the sternum.

Name the 4 ligaments of the sternoclavicular joint and describe their functions.

1. **Anterior sternoclavicular ligament:** Reinforces the capsule anteriorly
2. **Posterior sternoclavicular ligament:** Reinforces the capsule posteriorly
3. **Interclavicular ligament:** Stabilizes the medial ends of the clavicles
4. **Costoclavicular ligament:** Attaches the inferior surface of the medial end of the clavicle with the first rib and its costal cartilage

Describe the acromioclavicular joint.

The acromioclavicular joint is a synovial joint that joins the lateral end of the clavicle with the acromion of the scapula.

Name the 2 ligaments of the acromioclavicular joint.

1. Acromioclavicular ligament
2. Coracoclavicular ligament (formed by the conoid and trapezoid ligaments)

What is a "separated" shoulder?

Shoulder "separation" is the informal term used to describe dislocation or subluxation (partial dislocation) of the acromioclavicular joint ("AC joint"), most often caused by a fall onto the shoulder. In contrast, the term "shoulder dislocation" generally refers to dislocation at the *glenohumeral* joint.

Describe the glenohumeral joint.

The glenohumeral joint is a ball-and-socket joint between the glenoid fossa of the scapula and the head of the humerus.

Name the 6 ligaments of the glenohumeral joint.

1. Superior glenohumeral ligament
2. Middle glenohumeral ligament
3. Inferior glenohumeral ligament
4. Transverse humeral ligament
5. Coracohumeral ligament
6. Coracoacromial ligament

The glenohumeral joint allows for which types of movements?

Flexion and extension, abduction and adduction, medial and lateral rotation, and circumduction

Name the 3 nerves that innervate the glenohumeral joint.

1. Axillary nerve
2. Suprascapular nerve
3. Lateral pectoral nerve

What nerve is most commonly affected with anterior shoulder dislocation?

Axillary nerve. Damage to this nerve occurs in up to 30% of cases and usually resolves over time without specific treatment.

Name the 3 bursae of the glenohumeral joint.

1. Subacromial bursa
2. Subdeltoid bursa
3. Subscapular bursa

What is the function of these bursae?

To reduce friction between the rotator cuff and the coracoacromial arch during movement of the glenohumeral joint.

Name the muscles involved in the following movements at the glenohumeral joint:

Adduction

Pectoralis major, latissimus dorsi, teres major, triceps, and subscapularis

Abduction

Deltoid and supraspinatus

Flexion

Pectoralis major, anterior part of the deltoid, coracobrachialis, and biceps brachii

Extension

Latissimus dorsi, posterior part of the deltoid, triceps, and teres major

Medial (internal) rotation

Subscapularis, pectoralis major, anterior part of the deltoid, latissimus dorsi, and teres major

Lateral (external) rotation

Infraspinatus, teres minor, and posterior part of the deltoid

AXILLA

What is the axilla?

The pyramidal area at the junction of the upper extremity and the trunk (underarm)

What are the boundaries of the axilla:

Medially?

Ribs 1–4, the intercostal muscles, and the serratus anterior muscle

Laterally?

The humerus (specifically, the floor of the intertubercular groove)

Anteriorly?

The pectoralis major and pectoralis minor muscles

Posteriorly?

The scapula and the subscapularis, teres major, and latissimus dorsi muscles

What forms the base of the axilla?

The axillary fascia and skin

The apex?

The interval between the clavicle, scapula, and first rib

What 6 structures are contained within the axilla?	1. Axillary artery 2. Axillary vein 3. Axillary lymph nodes 4. Branches of the brachial plexus 5. Biceps brachii muscle (the long and short heads) 6. Coracobrachialis muscle
What is the axillary sheath?	A continuation of the cervical fascia into the axilla that encloses the axillary artery, the axillary vein, and the brachial plexus

VASCULATURE

Axillary Artery

Describe the origin and fate of the axillary artery.	The subclavian artery becomes the axillary artery at the lateral border of the first rib. The axillary artery becomes the brachial artery at the inferior border of the teres major muscle.
Delineate the 3 parts of the axillary artery, and name the branches from each. (Note that the first part has 1 branch, the second part 2, and the third part 3!)	
First part?	Extends from the lateral border of the first rib to the superior border of the pectoralis minor muscle, giving off the superior thoracic artery
Second part?	Extends deep to the pectoralis minor muscle, giving off the thoracoacromial artery and the lateral thoracic artery
Third part?	Extends from the inferior border of the pectoralis minor muscle to the inferior border of the teres major muscle, giving off the subscapular, anterior circumflex humeral, and posterior circumflex humeral arteries
What are the 2 branches of the subscapular artery?	1. Circumflex scapular artery 2. Thoracodorsal artery

Which 2 branches of the axillary artery anastomose with one another on the surgical neck of the humerus?

The anterior and posterior circumflex humeral arteries

Which artery provides the majority of the blood supply to the humeral head?

The anterior circumflex humeral artery gives off an ascending branch that supplies the majority of the blood supply to the humeral head.

Identify the labeled arteries on the following figure:

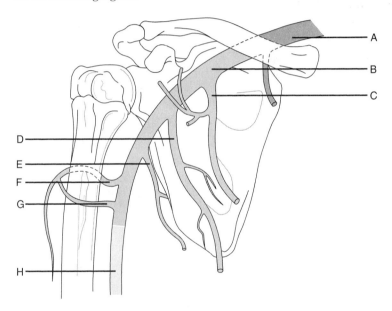

A = Subclavian artery
B = Axillary artery
C = Thoracoacromial artery
D = Lateral thoracic artery
E = Subscapular artery
F = Posterior circumflex humeral artery
G = Anterior circumflex humeral artery
H = Brachial artery

Note that the superior thoracic artery branches off the first part of the axillary posteriorly and therefore is not visible in this view.

Axillary Vein

Describe the origin and fate of the axillary vein.

The axillary vein originates at the inferior border of the teres major muscle as a continuation of the basilic vein and becomes the subclavian vein at the lateral border of the first rib. Note that the borders of the axillary vein parallel those of the axillary artery.

Within the axilla, what is the relationship between the axillary vein and artery?

The vein lies superficial (interior) to the artery.

Axillary Lymph Nodes

Name the 5 groups of axillary lymph nodes.

1. Central
2. Lateral
3. Subscapular (posterior)
4. Pectoral
5. Apical

What is the course of lymphatic drainage in the axillary region?

The lateral, pectoral, and subscapular (posterior) nodes drain into the central nodes, which in turn drain into the apical nodes. The apical nodes drain into the subclavian trunks.

BRACHIAL PLEXUS

What is the brachial plexus?

A large network of nerves that originates in the neck and extends into the axilla, giving rise to most of the nerves that supply the upper extremity

What are the 5 segmental branchings of the brachial plexus?

Ron **T**aylor **D**rinks **C**old **B**eer
Rami
Trunks
Divisions
Cords
Branches (terminal)

The ventral primary rami of which spinal cord segments contribute to the brachial plexus?

C5, C6, C7, C8, and T1

The rami leading to the brachial plexus run between which 2 muscles?

The anterior and middle scalene muscles

Which 2 nerves branch off from the rami of the brachial plexus before the rami become trunks?

1. **The dorsal scapular nerve:** From the C5 rami, supplies motor to the rhomboid minor and rhomboid major muscles
2. **The long thoracic nerve:** From the C5, C6, and C7 rami, supplies motor to the serratus anterior muscle (remember scapular winging? (Note that neither of these nerves has a sensory component.)

Which rami contribute to which trunks?

Superior trunk: Formed from the rami of C5 and C6
Middle trunk: Continuation of the ramus of C7
Inferior trunk: Formed from the rami of C8 and T1

Which 2 nerves branch off the superior trunk of the brachial plexus?

1. **The suprascapular nerve:** From the C5 and C6 rami, supplies motor to the supraspinatus and infraspinatus muscles and sensation to the glenohumeral joint, runs under the superior transverse scapular ligament (Remember: The **N**erve is like the **N**avy!)
2. **The nerve to subclavius:** From the C5 ramus, supplies motor to the subclavius muscle (no sensory contribution)

What do the trunks divide into?

Each trunk splits into an anterior division and a posterior division.

How are the cords formed from the anterior and posterior divisions of the trunks?

Lateral cord: Formed from the anterior divisions of the superior and middle trunks
Posterior cord: Formed from the posterior divisions of all 3 trunks
Medial cord: Continuation of the anterior division of the inferior trunk

The lateral cord gives rise to which branch?

The lateral pectoral nerve: From the C5, C6, and C7 rami, supplies motor to the pectoralis major muscle and contributes to the motor innervation of the pectoralis minor muscle

What are the 2 terminal branches of the lateral cord?

1. **The musculocutaneous nerve:** From the C5, C6, and C7 rami, supplies motor to the anterior compartment of the arm (the coracobrachialis, biceps brachii, and brachialis muscles); terminates as the lateral cutaneous nerve of the forearm to provide sensation to the lateral aspect of the forearm.
2. The lateral cord contribution to the median nerve

The medial cord gives rise to which 3 branches?

1. **The medial pectoral nerve:** From the C8 and T1 rami, supplies motor to the pectoralis minor and pectoralis major muscles; no sensory contribution
2. **The medial brachial cutaneous nerve:** From the C8 and T1 rami, supplies sensation to the skin on the medial aspect of the arm; no motor contribution
3. **The medial antebrachial cutaneous nerve:** From the C8 and T1 rami, supplies sensation to the skin on the medial side of the forearm; no motor contribution

Where does the medial pectoral nerve lie in relation to the lateral pectoral nerve?

Lateral to it! The medial pectoral nerve is named "medial" because it arises from the medial cord of the brachial plexus, not because of its position relative to the lateral pectoral nerve. Similarly, the lateral pectoral nerve is designated "lateral" because it arises from the lateral cord; it lies medial to the medial pectoral nerve.

What are the 2 terminal branches of the medial cord?

1. The ulnar nerve
2. The medial cord contribution of the median nerve

Which terminal nerve branch of the brachial plexus receives contributions from both the medial and lateral cords?

The median nerve: From the C5, C6, C7, C8, and T1 rami

The posterior cord gives rise to which 3 branches?

1. **The upper subscapular nerve:** From the C5 and C6 rami, provides motor innervation to the upper portion of the subscapularis muscle; no sensory contribution
2. **The thoracodorsal nerve:** From the C7 and C8 rami, provides motor innervation to the latissimus dorsi muscle; no sensory contribution
3. **The lower subscapular nerve:** From the C5 and C6 rami, provides motor innervation to the lower portion of the subscapularis muscle as well as the teres major muscle; no sensory contribution

Identify the labeled nerves on the following diagram of the brachial plexus: (Come on! You can do it!)

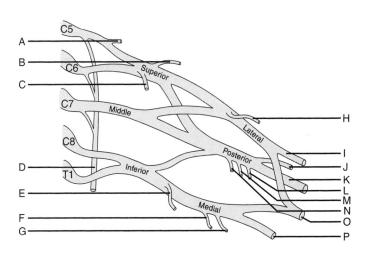

A = Dorsal scapular nerve
B = Suprascapular nerve
C = Nerve to subclavius
D = Long thoracic nerve
E = Medial pectoral nerve
F = Medial brachial cutaneous nerve
G = Medial antebrachial cutaneous erve
H = Lateral pectoral nerve
I = Musculocutaneous nerve
J = Axillary nerve
K = Radial nerve
L = Lower subscapular nerve
M = Thoracodorsal nerve
N = Upper subscapular nerve
O = Median nerve
P = Ulnar nerve

Note: The brachial plexus is a key anatomic structure that will come up again and again during your anatomy course and beyond; you would be well served to memorize it so that you can reproduce a diagram of it on command.

What are the 2 terminal branches of the posterior cord?

1. **The axillary nerve:** From the C5 and C6 rami, innervates the deltoid and teres minor muscles and eventually becomes the lateral brachial cutaneous nerve, which supplies the skin on the lateral arm
2. **The radial nerve:** From the C5, C6, C7, C8, and T1 rami

Which nerve is the largest branch of the brachial plexus?

The radial nerve

Which muscles are innervated by the radial nerve?

The extensors of the upper limb

ARM (BRACHIUM)

MUSCLES OF THE ARM

What divides the arm into anterior fascial (flexor) and posterior fascial (extensor) compartments?

The medial and lateral intermuscular septa and the humerus

Which muscles of the brachium lie in the:

Anterior fascial (flexor) compartment?

BBC
Brachialis muscle
Biceps brachii muscle
Coracobrachialis muscle

Posterior fascial (extensor) compartment?

Triceps brachii muscle

Brachialis Muscle

Origin?

The distal half of the anterior surface of the humerus

Insertion?

The ulnar tuberosity and the coronoid process of the ulna

Innervation?

The musculocutaneous nerve

Action?

Main flexor of the forearm at the humeroulnar (elbow) joint

Biceps Brachii Muscle

Origin?

Long head: The supraglenoid tubercle of the scapula
Short head: The coracoid process of the scapula

Insertion?

The radial tuberosity and the fascia of the medial forearm via the bicipital aponeurosis

Innervation?

The musculocutaneous nerve

Action?

Supination of the radioulnar joints and flexion of the humeroulnar joint from a supine position

Describe the course of the tendon of the long head of the biceps brachii muscle.

The tendon of the long head of the biceps brachii muscle crosses the distal humerus (enclosed in a fibrous capsule) and descends in the intertubercular groove to the radial tuberosity.

Coracobrachialis Muscle

Origin?

The tip of the coracoid process of the scapula

Insertion?

The medial surface of the humerus, about halfway down

Innervation?

The musculocutaneous nerve (pierces the body of the muscle as it travels distally)

Action?

Assists in flexion and adduction of the arm at the glenohumeral joint

Triceps Brachii Muscle

Origin?

Long head: The infraglenoid tubercle of the scapula

Lateral head: The posterior surface of the humerus, proximal (superior) to the radial groove

Medial head: The posterior surface of humerus, distal (inferior) to the radial groove

Insertion?

The olecranon of the ulna

Innervation?

The radial nerve (Remember: The radial nerve innervates **all** of the extensors in the upper limb.)

Action?

The triceps brachii is the chief extensor of the forearm at the humeroulnar joint; in addition, the long head of the triceps brachii steadies the head of the abducted humerus.

Which small muscle assists the triceps brachii muscle in extending the forearm?

The anconeus muscle, located in the forearm

VASCULATURE OF THE ARM

Arteries

Which artery represents the principal arterial supply to the upper limb?

The brachial artery

Identify the labeled arteries on the following figure of the arm and forearm:

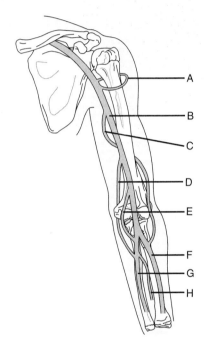

A = Anterior and posterior circumflex humeral arteries
B = Brachial artery
C = Deep brachial artery
D = Superior ulnar collateral artery
E = Inferior ulnar collateral artery
F = Radial artery
G = Ulnar artery
H = Anterior interosseus artery

Describe the course of the brachial artery.

The brachial artery originates at the inferior border of the teres major muscle as a continuation of the axillary artery and courses distally with the median nerve in the "bicipital groove" anterior to the medial intermuscular septum before ending in the cubital fossa by dividing into the ulnar and radial arteries.

The superior ulnar collateral artery pierces which fascial membrane?

The medial intermuscular septum

The superior ulnar collateral artery travels with which nerve behind the medial epicondyle?

The ulnar nerve

Veins

Name the 2 main superficial veins of the arm.

The cephalic vein and the basilic vein

Identify the labeled superficial veins on the following figure of the arm and forearm:

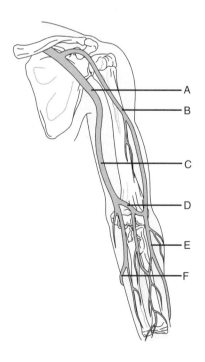

A = Axillary vein
B = Cephalic vein
C = Basilic vein
D = Median cubital vein
E = Cephalic vein
F = Basilic vein

Where do the cephalic and basilic veins originate from?

The dorsal venous arch of the hand

Which superficial vein is more lateral?

The cephalic vein

Which superficial vein runs through the deltopectoral triangle?

The cephalic vein

Which vein forms a communication between the basilic and cephalic veins in the cubital fossa?

The median cubital vein

INNERVATION OF THE ARM

Which 4 major nerves run through the arm?

1. Median nerve
2. Ulnar nerve
3. Radial nerve
4. Musculocutaneous nerve

Which of these do not branch in the arm?

The median and ulnar nerves

The median nerve lies in close proximity to which artery?

The brachial artery (The median nerve is located just anterior to it.)

Describe the course of the radial nerve in the arm.

The radial nerve enters the arm posterior to the brachial artery and medial to the humerus. It passes inferolaterally along the humerus in the radial groove (accompanied by the deep brachial artery), and pierces the lateral intermuscular septum to run between the brachialis and brachioradialis muscles. It then runs anterior to the lateral epicondyle of the humerus and divides into deep and superficial branches just distal to the lateral epicondyle.

Describe the course of the musculocutaneous nerve.

The musculocutaneous nerve pierces the coracobrachialis muscle and descends between the biceps brachii and brachialis muscles. Eventually, the musculo-cutaneous nerve becomes the lateral antebrachial cutaneous nerve. (Recall that the medial antebrachial cutaneous nerve is a branch from the medial cord of the brachial plexus.)

ELBOW REGION

CUBITAL FOSSA

What is the cubital fossa?

The hollow area on the anterior surface of the elbow

What forms the floor of the cubital fossa?

The brachialis and supinator muscles

The roof?

The superficial fascia, skin, and bicipital aponeurosis

List 4 key structures found in the cubital fossa.

1. The brachial artery
2. The median nerve
3. The radial artery
4. The ulnar artery

Where does the median cubital vein lie in relation to the bicipital aponeurosis?

Superficial to it. The veins in the region of the cubital fossa are a favorite target for phlebotomists because of their superficial position.

What are the boundaries of the cubital fossa:
    Laterally?

The brachioradialis muscle (a superficial posterior muscle in the forearm)

    Medially?

The pronator teres muscle (a superficial anterior muscle in the forearm)

Superiorly?   Imagine a line between the medial and
                                   lateral epicondyles of the humerus

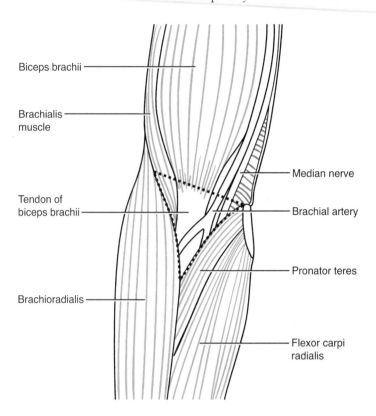

Biceps brachii

Brachialis
muscle

Tendon of
biceps brachii

Brachioradialis

Median nerve

Brachial artery

Pronator teres

Flexor carpi
radialis

HUMEROULNAR (ELBOW) JOINT

**Describe the humeroulnar
joint.**

The humeroulnar joint is a synovial joint
in which the trochlea of the humerus
articulates with the trochlear notch of the
ulna, and the capitulum of the humerus
articulates with the head of the radius.

**Name the muscles involved
in:**

**Flexion of the humeroul-
nar joint?**

The brachialis, biceps brachii,
brachioradialis, and pronator teres
muscles

Extension of the humer-oulnar joint?	The triceps brachii and anconeus muscles
Name the 3 ligaments that reinforce the humeroulnar joint.	1. Annular ligament 2. Radial collateral ligament 3. Ulnar collateral ligament
Describe the annular ligament.	The annular ligament travels around most of the head of the radius, preventing withdrawal of the head from its socket. The annular ligament then fuses with the radial collateral ligament.

FOREARM (ANTEBRACHIUM)

MUSCLES OF THE FOREARM

What are the 2 groups of forearm muscles and on which side of the forearm are they generally located?	1. The medial flexor-pronator (anterior forearm) group 2. The lateral extensor-supinator (posterior forearm) group

Anterior Forearm Muscles

Classify the 8 flexor-pronator muscles by action. Which flexor-pronator muscles:	
Pronate the forearm and hand?	1. Pronator teres muscle 2. Pronator quadratus muscle
Flex the hand at the radiocarpal joint?	1. Flexor carpi radialis muscle 2. Flexor carpi ulnaris muscle 3. Palmaris longus muscle
Flex the interphalangeal joints of the digits?	1. Flexor digitorum superficialis muscle 2. Flexor digitorum profundus muscle 3. Flexor pollicis longus muscle
What is the only muscle that can flex the distal interphalangeal joints?	The flexor digitorum profundus muscle (This muscle also assists in flexion at the metacarpophalangeal and wrist joints.)
Which flexor-pronator muscles form the:	
Superficial layer of anterior forearm muscles?	From lateral to medial: 1. The pronator teres muscle 2. The flexor carpi radialis muscle

3. The palmaris longus muscle
4. The flexor carpi ulnaris muscle

Intermediate layer of anterior forearm muscles?

The flexor digitorum superficialis muscle

Deep layer of anterior muscle?

1. The flexor digitorum profundus forearm muscles
2. The flexor pollicis longus muscle
3. The pronator quadratus muscle

What is the common origin of the flexor-pronator muscle group?

The common flexor tendon, from the medial humeral epicondyle (the "funny bone")

Identify the superficial flexor-pronator muscles on the following figure:

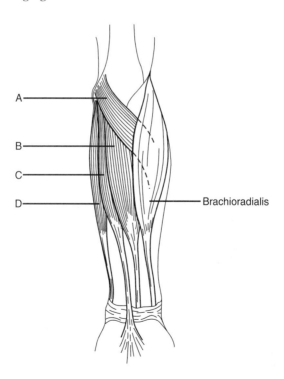

A = Pronator teres muscle
B = Flexor carpi radialis muscle
C = Palmaris longus muscle
D = Flexor carpi ulnaris muscle

Which 3 flexor-pronator muscles do not share this common origin?

The deep flexor-pronator muscles (i.e., the flexor digitorum profundus, the flexor pollicis longus, and the pronator quadratus muscles) do not arise from the common flexor tendon.

Which of the superficial flexor muscles is congenitally absent in approximately 10% of people and what is the clinical significance of this?

The palmaris longus. It is not essential for proper function and therefore is often used as a graft for tendon transfers.

Identify the deep flexor-pronator muscles on the following figure:

A = Flexor digitorum profundus muscle
B = Flexor pollicis longus muscle
C = Pronator quadratus muscle

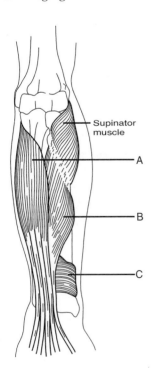

Supinator muscle

A

B

C

Pronator Teres Muscle

Origin?   The medial epicondyle of the humerus

Insertion?   The lateral surface of the radius, about
                                               halfway down

Innervation?   The median nerve

Action?   Pronates the hand and forearm and
                                               assists in flexion of the forearm

Flexor Carpi Radialis Muscle

Origin?   The medial epicondyle of the humerus

Insertion?   The bases of the second and third
                                               metacarpal bones

Innervation?   The median nerve

Action?   Flexes the wrist and abducts (radially
                                               deviates) the wrist/hand

**What artery runs   Radial artery
immediately lateral to the
flexor carpi radialis?**

**The tendon of the flexor   The trapezium (i.e., the first bone in the
carpi radialis muscle runs   second row of carpal bones)
through a vertical groove in
which carpal bone?**

Palmaris Longus Muscle

Origin?   The medial epicondyle of the humerus

Insertion?   The flexor retinaculum (i.e., a thickened
                                               portion of the antebrachial fascia) and
                                               the palmar aponeurosis (i.e., a thickened
                                               portion of the palmar fascia)

Innervation?   The median nerve

Action?   Flexes the hand and forearm

Flexor Carpi Ulnaris Muscle

Origin?

The medial epicondyle of the humerus, the medial olecranon of the ulna, and the posterior border of the ulna

Insertion?

The pisiform bone, the hook of the hamate, and the base of the fifth metacarpal bone

Innervation?

The ulnar nerve

Action?

Flexes the wrist and adducts (ulnar deviates) the wrist and hande

Flexor Digitorum Superficialis Muscle

Origin?

The medial epicondyle of the humerus, the coronoid process of the ulna, and the oblique line of the proximal anterior radius

Insertion?

The middle phalanges of the digits (not the thumb)

Innervation?

The median nerve

Action?

Flexes the proximal interphalangeal joints of the medial 4 digits (can assist in flexing the hand and metacarpophalangeal joint)

Flexor Digitorum Profundus Muscle

Origin?

The anteromedial surface of the ulna and the interosseus membrane

Insertion?

The bases of the distal phalanges of the digits (not thumb)

Innervation?

The ulnar nerve (ring and small fingers) and the anterior interosseous branch of the median nerve (index and middle fingers)

Action?

Flexes the distal interphalangeal joints and assists in flexing the metacarpo-phalangeal joints and radiocarpal (wrist) joint

What is the flexor tendon sheath?

The fibroosseous tunnel lined with tenosynovium on the palmar aspect of the digits that contains the flexor digitorum profundus and sublimis tendons

Flexor Pollicis Longus Muscle

Origin?

The anterior surface of the radius, the interosseous membrane, and the coronoid process of the ulna

Insertion?

The base of the distal phalanx of the thumb

Innervation?

The anterior interosseous branch of the median nerve

Action?

Flexes the interphalangeal joint of the thumb

Pronator Quadratus Muscle

Origin?

The anterior surface of the distal ulna

Insertion?

The anterior surface of the distal radius

Innervation?

The anterior interosseous branch of the median nerve

Action?

Pronation of the forearm

Posterior Forearm Muscles

Classify the 9 extensor-supinator muscles by action. Which extensor-supinator muscles:

Extend the radiocarpal joint?

1. The extensor carpi radialis longus muscle
2. The extensor carpi radialis brevis muscle
3. The extensor carpi ulnaris muscle

Extend the metacarpophalangeal joints of the digits?

1. The extensor digitorum muscle
2. The extensor indicis muscle
3. The extensor digiti minimi muscle

Extend the thumb?

1. The abductor pollicis longus muscle
2. The extensor pollicis brevis muscle
3. The extensor pollicis longus muscle

Which structure prevents "bowstringing" of the extensor tendons of the wrist when the hand is hyperextended at the carpal joint?

The extensor retinaculum (i.e., a thickening of the deep fascia of the forearm at the wrist)

What are the extensor expansions?

Located on the distal ends of the metacarpal bones and on the phalanges, the extensor expansions are formed when the extensor tendons flatten out over the bone.

Identify the labeled muscles on the following diagram of the superficial muscles of the posterior forearm:

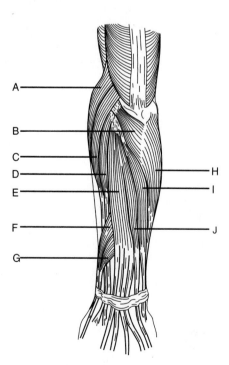

A = Brachioradialis muscle
B = Anconeus muscle
C = Extensor carpi radialis longus
    muscle
D = Extensor carpi radialis brevis muscle
E = Extensor digitorum muscle
F = Abductor pollicis longus muscle
G = Extensor pollicis brevis muscle
H = Flexor carpi ulnaris muscle
I = Extensor carpi ulnaris muscle
J = Extensor digiti minimi

Identify the labeled muscles on the following diagram of the deep muscles of the posterior forearm:

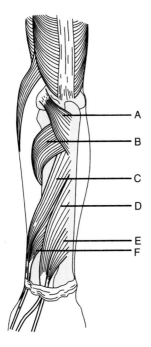

A = Anconeus muscle
B = Supinator muscle
C = Abductor pollicis longus muscle
D = Extensor pollicis longus muscle
E = Extensor indicis muscle
F = Extensor pollicis brevis muscle

What is the origin of the deep muscles of the posterior forearm?

The interosseous membrane and the posterior surfaces of the ulna and radius

Which nerve innervates all the muscles of the posterior forearm?

The radial nerve, which has superficial and deep branches. The deep branch becomes the posterior interosseous nerve.

Anconeus Muscle

Origin?

The lateral epicondyle of the humerus (posteriorly)

Insertion?

The lateral surface of the olecranon and the superior part of the posterior surface of the ulna

Innervation?

The radial nerve

Action?

Assists the triceps brachii in extending the forearm at the humeroulnar joint

Brachioradialis Muscle

Origin?

The proximal two thirds of the lateral supracondylar ridge of the humerus

Insertion?

The lateral surface of the distal radius

Innervation?

The radial nerve

Action?

Flexes the forearm at the humeroulnar (elbow) joint

Extensor Carpi Radialis Brevis Muscle

Origin?

The lateral epicondyle of the humerus

Insertion?

The base of the third metacarpal bone

Innervation?

The radial nerve

Action?

Extends the wrist and abducts (radially deviates) the hand at the radiocarpal joint

Extensor Carpi Radialis Longus Muscle

Origin?

The lateral supracondylar ridge of the humerus

Insertion?

The base of the second metacarpal bone

Innervation?	The radial nerve
Action?	Extends the wrist and abducts (radially deviates) the hand at the radiocarpal joint

Extensor Carpi Ulnaris Muscle

Origin?	The lateral epicondyle and posterior surface of the ulna
Insertion?	The base of the fifth metacarpal bone
Innervation?	The radial nerve
Action?	Extends and adducts (ulnarly deviates) the hand at the radiocarpal joint

Extensor Digiti Minimi Muscle

Origin?	The common extensor tendon and the interosseus membrane
Insertion?	The extensor expansions and the dorsal bases of the middle and distal phalanges
Innervation?	The radial nerve (posterior interosseous nerve branch)
Action?	Extends the fifth digit (i.e., the "little" finger)

Extensor Digitorum Muscle

Origin?	The lateral epicondyle of the humerus
Insertion?	The extensor expansions and the dorsal bases of the middle and distal phalanges
Innervation?	The radial nerve (posterior interosseous branch)
Action?	Extends the fingers at the metacarpopha-langeal joints and assists with extension of the hand at the radiocarpal joint

Abductor Pollicis Longus Muscle

Origin?

The interosseus membrane and the posterior surfaces of the radius and ulna

Insertion?

The base of the first metacarpal bone

Innervation?

The radial nerve (posterior interosseous branch)

Action?

Abducts the thumb at the carpometacarpal joint and the hand at the radiocarpal joint

Extensor Pollicis Brevis Muscle

Origin?

The interosseus membrane and posterior surface of the radius

Insertion?

The base of the proximal phalanx of the thumb

Innervation?

The radial nerve (posterior interosseous branch)

Action?

Extends the proximal phalanx of the thumb and abducts the hand at the radiocarpal joint

Extensor Pollicis Longus Muscle

Origin?

The interosseus membrane and the posterior surface of the ulna

Insertion?

The base of the distal phalanx of the thumb

Innervation?

The radial nerve (posterior interosseous branch)

Action?

Extends the distal phalanx of the thumb

Extensor Indicis Muscle

Origin?

The interosseus membrane and the posterior surface of the ulna

Insertion?	The extensor expansion of the second digit (the index finger)
Innervation?	The radial nerve (posterior interosseous branch)
Action?	Extends the index finger

Supinator Muscle

Origin?	The lateral epicondyle of the humerus, the radial collateral and annular ligaments, the supinator fossa, and the crest of the ulna
Insertion?	The lateral, posterior, and anterior surfaces of the proximal third of the radius
Innervation?	The radial nerve
Action?	Supination of the forearm and hand at the radioulnar joint

VASCULATURE OF THE FOREARM

Which artery in the forearm is often used to palpate an arterial pulse?	The radial artery, which lies lateral to the tendon of the flexor carpi radialis muscle
The ulnar artery accompanies which nerve on its course between the 2 heads of the flexor digitorum superficialis muscle?	The median nerve (Note the ulnar artery subsequently runs medially to join the ulnar nerve.)
The ulnar artery lies between which 2 muscles?	The flexor digitorum superficialis and the flexor digitorum profundus
The common interosseous artery (a branch of the ulnar artery) divides into which 2 arteries?	The anterior interosseous artery and the posterior interosseous artery

Which artery runs with the superficial branch of the radial nerve under the brachioradialis muscle?

Radial artery

What clinical deficit results with complete occlusion of either the radial or ulnar artery?

No deficit. There is adequate "collateral flow" to prevent ischemia of the forearm or hand.

INNERVATION OF THE FOREARM

Describe the course of the ulnar nerve at the elbow.

The ulnar nerve passes between the medial epicondyle of the humerus (the "funny bone") and the olecranon of the ulna to enter the forearm. Compression of the ulnar nerve at this level may lead to numbness or weakness of the hand in an ulnar nerve distribution (cubital tunnel syndrome).

The following nerves travel between the heads of which flexor-pronator muscles:
    Median nerve?

The pronator teres muscle

    Ulnar nerve?

The flexor carpi ulnaris muscle

JOINTS AND LIGAMENTS OF THE FOREARM

Describe the proximal radioulnar joint.

The proximal radioulnar joint is a pivot-type synovial joint in which the head of the radius articulates with the radial notch of the ulna.

Describe the distal radioulnar joint.

The distal radioulnar joint is a pivot-type synovial joint in which the head of the radius ulna articulates with the ulnar notch of the ulna.

The proximal and distal radioulnar joints allow for which movements of the forearm?

Pronation and supination

Which muscles are involved in:

Pronation of the forearm?

The pronator quadratus and pronator teres muscles

Supination of the forearm?

The supinator and biceps brachii muscles

WRIST (CARPUS)

Describe the radiocarpal (wrist) joint.

The radiocarpal joint is a condyloid joint where the radius and the triangular fibrocartilage complex articulate with the scaphoid, lunate, and triquetral bones.

Which movements are possible at the radiocarpal joint?

Flexion and extension, abduction (radial deviation) adduction (ulnar deviation), and circumduction

Name the ligaments associated with the radiocarpal joint.

1. Dorsal radioulnar ligament
2. Palmar radioulnar ligament
3. Dorsal ulnocarpal ligament
4. Palmar ulnocarpal ligament
5. Dorsal radiocarpal ligament
6. Palmar radiocarpal ligament
7. Radial collateral ligament
8. Ulnar collateral ligament

Which movements are allowed between the proximal and distal carpal rows?

Flexion and abduction of the hand

Which structure converts the carpal groove into a "tunnel"?

The flexor retinaculum

How is the flexor retinaculum of the wrist formed?

The deep fascia of the forearm thickens anteriorly at the wrist.

Describe the attachments of the flexor retinaculum.

The scaphoid and trapezium (i.e., the lateral-most carpal bones) laterally and the pisiform and hamate (the medial-most carpal bones) medially

Name the structures that enter the palm superficial to the flexor retinaculum.

UP UP above the flexor retinaculum
Ulnar nerve
Palmaris longus tendon
Ulnar artery
Palmar cutaneous branch of the median nerve

Which nerve wraps around the hook of the hamate bone as it travels distally and what is the clinical significance of this?

The deep branch of the ulnar nerve. Fracture of the hook of the hamate can cause ulnar neuropathy in the hand (weakness of the interossei muscles and ulnar-sided lumbricals).

What are the anterior and posterior boundaries of the carpal tunnel?

The flexor retinaculum anteriorly and the carpal bones posteriorly. Surgical release of the carpal tunnel involves division of the flexor retinaculum.

What structures pass through the carpal tunnel?

1. Median nerve
2. Flexor pollicis longus tendon
3. Flexor digitorum profundus tendons (4)
4. Flexor digitorum superficialis tendons (4)

What is the clinical significance of compression within the carpal tunnel?

Compression within the carpal tunnel (carpal tunnel syndrome) causes median neuropathy, which can manifest as wrist pain and weakness with atrophy of the muscles of the hand innervated by the median nerve (thenar atrophy).

What structure, deep to the flexor retinaculum, encloses the flexor digitorum profundus and flexor digitalis superficialis tendons?

The common flexor synovial sheath

How is the extensor retinaculum formed?

The deep fascia of the forearm thickens posteriorly at the wrist.

Describe the attachments of the extensor retinaculum.	The styloid process of the ulna medially, and the triquetrum and pisiform bones laterally

HAND (MANUS)

ANATOMIC SNUFF BOX

What are the boundaries of the anatomic snuff box:	
Posteriorly?	The tendon of the extensor pollicis longus muscle
Anteriorly?	The tendons of the extensor pollicis brevis and abductor pollicis longus muscles
Proximally?	The styloid process of the radius
What forms the floor of the anatomic snuff box?	The scaphoid and trapezium bones
Which artery lies in the anatomic snuff box?	The radial artery
Tenderness over the anatomic snuff box suggests fracture of which bone?	The scaphoid bone

JOINTS AND LIGAMENTS OF THE HAND

Which ligaments support the metacarpophalangeal joints?	Each joint is supported by 1 palmar ligament and 2 collateral ligaments.
What is the clinical name for enlargement of the:	
Proximal interphalangeal joints?	Bouchard's nodes (characteristic in rheumatoid arthritis)
Distal interphalangeal joints?	Heberden's nodes (characteristic in osteoarthritis)

What structures keep the flexor tendons from bowstringing at the fingers?

Flexor pulleys on the volar aspect of the phalanges and metacarpal heads. These structures are involved in "trigger finger" (stenosing tenosynovitis), a condition where nodules on the flexor tendons catch on these pulleys within the flexor tendon sheath, causing the fingers to lock in a flexed position.

FASCIAE AND MUSCLES OF THE HAND

Name the 4 fascial compartments in the hand.

1. Thenar compartment
2. Adductor compartment
3. Hypothenar compartment
4. Central compartment

What is the fibrous structure that overlays the tendons in the palm?

The palmar aponeurosis

Palmaris Brevis Muscle

Origin?

The transverse carpal ligament

Insertion?

The skin of the medial palm

Innervation?

The ulnar nerve

Action?

Wrinkles the skin of the medial palm

Intrinsic Hand Muscles

Name the 4 groups of intrinsic muscles of the hand.

1. Thenar muscles
2. Adductor pollicis muscle
3. Hypothenar muscles
4. Short muscles (i.e., the lumbricals and interossei)

Thenar Muscles

Name the 3 thenar muscles.

1. Abductor pollicis brevis muscle
2. Flexor pollicis brevis muscle
3. Opponens pollicis muscle

What is the major action of the thenar muscles?	Abduction, flexion, and opposition of the carpometacarpal joint of the thumb
Which nerve innervates the thenar muscles?	The median nerve

Abductor Pollicis Brevis Muscle

Origin?	The flexor retinaculum, the scaphoid bone, and the trapezium bone
Insertion?	The base of the proximal phalanx of the thumb
Action?	Abducts the thumb

Flexor Pollicis Brevis Muscle

Origin?	The flexor retinaculum and the trapezium bone
Insertion?	The base of the proximal phalanx of the thumb
Action?	Flexes the thumb

Opponens Pollicis Muscle

Origin?	The flexor retinaculum and the trapezium bone
Insertion?	The first metacarpal
Action?	Opposes the thumb to the other digits and rotates is medially

Adductor Pollicis Muscle

Origin?	2 heads: 1. The capitate bone, the bases of the second and third metacarpals 2. The palmar surface of the third metacarpal shaft
Insertion?	The proximal phalanx of the thumb

Innervation?	The ulnar nerve
Action?	Adducts the thumb
What vascular structure passes through the 2 heads of the adductor pollicis muscle?	The terminal branch of the radial artery as it forms the deep palmar arch

Hypothenar Muscles

Name the 3 hypothenar muscles.	1. Abductor digiti minimi muscle 2. Flexor digiti minimi brevis muscle 3. Opponens digiti minimi muscle
Which nerve innervates the hypothenar muscles?	The ulnar nerve

Abductor Digiti Minimi Muscle

Origin?	The pisiform bone and the tendon of the flexor carpi ulnaris muscle
Insertion?	Base of the proximal phalanx of the fifth digit
Action?	Abducts the fifth digit

Flexor Digiti Minimi Brevis Muscle

Origin?	The transverse carpal ligament and the hook of the hamate
Insertion?	Base of the proximal phalanx of the fifth digit
Action?	Flexes the proximal phalanx of the fifth digit

Opponens Digiti Minimi Muscle

Origin?	The transverse carpal ligament and the hook of the hamate
Insertion?	Medial side of the fifth metacarpal
Action?	Opposes the fifth digit and rotates it laterally

Short Muscles

What are the short muscles of the hand?	1. The lumbricals 2. The interossei (dorsal and palmar)

Lumbricals

Origin?	The lateral sides of the tendons of the flexor digitorum profundus muscle
Insertion?	The lateral sides of the extensor expansions
Innervation?	**Median nerve:** 2 radial-sided lumbricals **Ulnar nerve:** 2 medial-sided lumbricals
Action?	Flexion of the metacarpophalangeal joints and extension of the proximal interphalangeal joints

Interossei

Origin?	**Dorsal interossei:** The sides of the metacarpals **Palmar interossei:** The medial side of the second metacarpal and the lateral sides of the fourth and fifth metacarpals
Insertion?	**Dorsal interossei:** The bases of the proximal phalanges and the extensor expansions **Palmar interossei:** The bases of the proximal phalanges and the extensor expansions
Innervation?	The ulnar nerve
What are the actions of the: **Dorsal interossei muscles?**	1. Abduction of the digits from the axial line (i.e., the third digit). Remember **DAB: D**orsal **Ab**ducts. 2. Flexion of the metacarpophalangeal joints 3. Extension of the interphalangeal joints

Palmar interossei muscles?

1. Adduction of the digits to the axial line. Remember **PAD: P**almar **Ad**ducts
2. Flexion of the metacarpophalangeal joints
3. Extension of the interphalangeal joints

VASCULATURE AND INNERVATION OF THE HAND

In the hand, the radial artery divides into which vessels?

The princeps pollicis artery and the terminal branch of the radial artery, which contributes to the deep palmar arch

The princeps pollicis artery divides into which vessels?

2 proper digital arteries of the thumb and the radialis indices artery (along the radial aspect of the index finger)

Identify each region on the following figure by the nerve that innervates it:

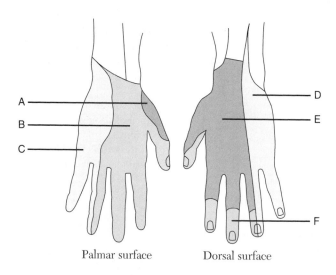

Palmar surface   Dorsal surface

A = Radial nerve
B = Median nerve
C = Ulnar nerve
D = Ulnar nerve
E = Radial nerve
F = Median nerve

Describe the clinical examination of sensory and motor function of the hand.

Sensory: Check sensation over:
　　Dorsal thumb (radial nerve)Palmar
　　thumb (median nerve)Little finger
　　(ulnar nerve)
Motor: Check:
　　Thumb apposition (median—
　　opponens pollicis)
　　Strength in spreading fingers (ulnar—
　　dorsal interossei)
　　Wrist extension (radial)

Note that radial nerve has no motor function within the hand!

POWER REVIEW

UPPER EXTREMITY

BONES

Name the 6 structures that attach to the coracoid process.

3 muscles (pectoralis minor insertion, short head of biceps origin, coracobrachialis origin) and 3 ligaments (coracoacromial, coracoclavicular, coracohumeral)

Which muscles originate just above and below the glenoid fossa?

The long head of the biceps brachii (supraglenoid tubercle) and the long head of the triceps brachii (infraglenoid tubercle), respectively

PECTORAL GIRDLE AND SHOULDER

Which 7 muscles pass from the scapula to the humerus and act on the shoulder joint?

1. Supraspinatus muscle
2. Infraspinatus muscle
3. Teres minor muscle
4. Subscapularis muscle
5. Deltoid muscle
6. Teres major muscles
7. Coracobrachialis

Which 5 muscles aid in the medial (internal) rotation of the humerus?

1. Deltoid muscle
2. Subscapularis muscle
3. Teres major muscle
4. Latissimus dorsi muscle
5. Pectoralis major muscle

Which 3 muscles rotate the humerus laterally (externally)?

1. Infraspinatus muscle
2. Teres minor muscle
3. Posterior part of the deltoid muscle

What are the 4 "rotator cuff" muscles?

SITS
Supraspinatus muscle
Infraspinatus muscle
Teres minor muscle
Subscapularis muscle

**Glenohumeral dislocation
usually occurs in which
direction? What nerve is
most commonly affected?**

Anterior; axillary nerve

**Which muscles insert on the:
    Greater tubercle of the
    humerus?**

The supraspinatus, infraspinatus, and
teres minor muscles

**Lesser tubercle of the
humerus?**

The subscapularis muscle

**Of the 4 rotator cuff
muscles, which does not
rotate the humerus?**

The supraspinatus muscle (The infraspinatus and teres minor muscles rotate
the humerus laterally [externally], while
the subscapularis muscle is involved in
medial [internal] rotation.)

**Identify the labeled
muscular attachments on the
following figure, showing
the posterior surfaces of the
clavicle, scapula, and
proximal humerus:**

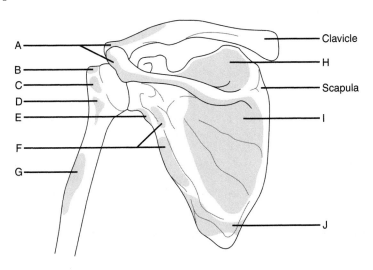

A = Origin of the deltoid muscle
B = Insertion of the supraspinatus muscle
C = Insertion of the infraspinatus muscle
D = Insertion of the teres minor muscle
E = Origin of the triceps muscle, long head
F = Origin of the teres minor muscle
G = Insertion of the deltoid muscle
H = Origin of the supraspinatus muscle
I = Origin of the infraspinatus muscle
J = Origin of the teres major muscle

AXILLA

What 6 structures are contained within the axilla?	1. Axillary artery 2. Axillary vein 3. Axillary lymph nodes 4. Branches of the brachial plexus 5. Biceps brachii muscle (the long and short heads) 6. Coracobrachialis muscle
The axillary artery extends from where to where?	The lateral border of the first rib to the inferior border of the teres major muscle
Name the branches from each of the 3 parts of the axillary artery (based on pectoralis minor muscle!):	
First part?	Superior thoracic artery
Second part?	Thoracoacromial artery, lateral thoracic artery
Third part?	Subscapular, anterior circumflex humeral, and posterior circumflex humeral arteries (Note that the first part has 1 branch, the second part 2, and the third part 3!)
What are the 5 segmental branchings of the brachial plexus?	**R**on **T**aylor **D**rinks **C**old **B**eer **R**ami **T**runks **D**ivisions **C**ords **B**ranches (terminal)

The ventral primary rami of which spinal cord segments contribute to the brachial plexus? C5, C6, C7, C8, and T1

Identify the labeled nerves on the following diagram of the brachial plexus: (Come on! You can do it!)

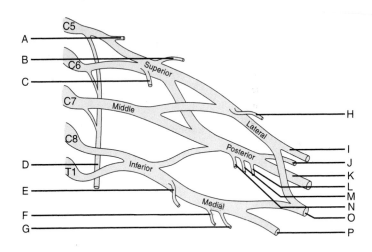

A = Dorsal scapular nerve
B = Suprascapular nerve
C = Nerve to subclavius
D = Long thoracic nerve
E = Medial pectoral nerve
F = Medial brachial cutaneous nerve
G = Medial antebrachial cutaneous nerve
H = Lateral pectoral nerve
I = Musculocutaneous nerve
J = Axillary nerve
K = Radial nerve
L = Lower subscapular nerve
M = Thoracodorsal nerve
N = Upper subscapular nerve
O = Median nerve
P = Ulnar nerve

The rami leading to the brachial plexus run between which 2 muscles?

The anterior and middle scalene muscles

What happens when the long thoracic nerve is injured?

Loss of serratus anterior function, resulting in "winging of the scapula"

ARM

What are the borders of the *quadrangular* space?

Superior: Teres minor
Inferior: Teres major
Medial: Long head of triceps
Lateral: Humerus

What structures pass through the quadrangular space?

The axillary nerve and posterior circumflex humeral artery

What are the borders of the *triangular* space?

Superior: Teres minor
Inferior: Teres major
Lateral: Long head of triceps

What structure passes through the triangular space?

Circumflex scapular artery

Name the 2 articular surfaces of the distal humerus.

Capitulum (lateral, articulates with radius)
Trochlea (medial, articulates with ulna)

Which nerve and vessel run in the spiral groove on the posterior humerus, and are therefore susceptible to damage following fracture of the humeral shaft?

The radial nerve and the deep brachial artery

Describe the muscles of the upper arm in terms of their fascial compartments.

Anterior fascial compartment: Biceps brachii, brachialis, and coracobrachialis muscles (flexors)
Posterior fascial compartment: Triceps brachii muscle (extensor)

Which nerve innervates the:

    Flexors of the upper limb?

The musculocutaneous nerve (innervates 3 "brachii" muscles—biceps brachii, brachialis, coracobrachialis)

    Extensors of the upper limb?

The radial nerve

Which muscle is the main flexor of the humeroulnar joint?

The brachialis muscle, **not** the biceps brachii muscle!

Which muscle is the main extensor of the humeroulnar joint?

The triceps brachii muscle

Weakened flexion of the humeroulnar joint and supination of the forearm suggests damage to which nerve?

The musculocutaneous nerve

ELBOW REGION AND FOREARM

Which nerve passes posterior to the medial humeral epicondyle ("funny bone")?

The ulnar nerve

All of the flexor muscles in the forearm are innervated by the median or ulnar nerves except for which 1?

The brachioradialis muscle. Despite being a flexor, the brachioradialis muscle is supplied by the radial nerve (the 1 exception to the rule that the radial nerve is the "nerve of the extensors").

Which nerve innervates all of the posterior muscles of the forearm?

The radial nerve

What is the origin of the deep muscles of the posterior forearm?

The interosseous membrane and the posterior surfaces of the ulna and radius

WRIST AND HAND

What carpal bone is the most commonly fractured?

The scaphoid bone. The fracture causes tenderness in the anatomic snuff box.

Which structures pass through the carpal tunnel?

The median nerve (this is the nerve that is compressed in carpel tunnel syndrome), the flexor pollicis longus tendon, the 4 flexor digitorum profundus tendons, and the 4 flexor digitorum superficialis tendons

Describe the course of the brachial artery.

Originates at the inferior border of teres major muscle as continuation of axillary artery and courses distally with median nerve in "bicipital groove" anterior to medial intermuscular septum before ending in cubital fossa by dividing into ulnar and radial arteries

The terminal branches of which arteries contribute to the superficial and deep palmar arches of the hand?

Terminal branch of radial artery—deep
Terminal branch of ulnar artery—superficial

The median nerve lies in close proximity to which artery?

The brachial artery (The median nerve is located just anterior to it.)

In the hand, which nerve innervates the:
    Thenar muscles?

Median nerve

    Hypothenar muscles?

Ulnar nerve

8   The Thorax

THORACIC WALL

Which bones form the skeleton of the thoracic wall?

1. Vertebrae T1–T12
2. Ribs 1–12 and their associated costal cartilages
3. The sternum

The sternum is composed of which 3 segments?

The manubrium, the body, and the xiphoid process

Which bones form the borders of the superior thoracic aperture?

Vertebra T1, rib 1, and the manubrium of the sternum

What is contained in the inferior to each rib?

Neurovascular bundles (**VAN: V**ein, costal grooves, immediately **A**rtery, **N**erve). Because the neurovascular bundle lies inferior to each rib, chest tube insertion (tube thoracostomy) is performed immediately *superior* to the rib.

How many intercostal spaces are there, and what do they contain?

11. Each contains a neurovascular bundle and 3 muscle layers.

Name the 3 intercostal muscle layers, from superficial to deep, and describe the continuation of each within the abdominal wall.

1. External intercostal (continuous with external oblique)
2. Internal intercostal (continuous with internal oblique)
3. Innermost intercostal (also continuous with internal oblique)

The anterior intercostal arteries are branches of which artery?

The internal thoracic (internal mammary) artery, the first branch of the subclavian artery

The posterior intercostal arteries are branches of which artery?

The descending thoracic aorta

Which muscle is found between the sternum and the ribs deep to the internal thoracic artery?

The transversus thoracis muscle (continuous with transversus abdominis muscle)

What are the terminal branches of the internal thoracic artery?

The superior epigastric artery and the musculophrenic artery

What is the chief surgical importance of the internal thoracic artery?

It is used for coronary artery bypass, where it is usually grafted to the left anterior descending (LAD) artery.

Which nerves innervate the chest wall?	The intercostal nerves (i.e., the ventral primary rami of spinal cord levels T1–T11) are both motor and sensory to the body wall and sensory to the parietal pleura and the periphery of the diaphragm.
Which dermatome supplies sensation to the:	
Nipple?	T4
Umbilicus?	T10
Name the muscles encountered during a lateral thoracotomy, from superficial to deep.	After an incision parallel to the ribs, the latissimus dorsi muscle is divided and the serratus anterior muscle is either divided or retracted anteriorly. In a posterolateral thoracotomy, the trapezius and rhomboid muscles are also usually divided. Access to the thoracic cavity is then obtained by dividing the 3 layers of intercostal muscles at the superior margin of the rib.

THORACIC CAVITY

The thoracic cavity is divided into how many compartments?	3—2 lateral compartments (which contain the lungs and pleurae) and a central compartment, known as the mediastinum, which contains the other thoracic structures

MEDIASTINUM

What are the boundaries of the mediastinum:	
Anteriorly?	The sternum
Posteriorly?	The thoracic vertebrae
Laterally?	The parietal pleura of the chest wall
Superiorly?	The superior thoracic aperture
Inferiorly?	The diaphragm

The superior and inferior mediastina are separated by a plane passing between which structures?	The sternal angle and the T4–T5 intervertebral disk

Superior Mediastinum

What structures are contained within the superior mediastinum?	1. The thymus gland 2. The trachea 3. The upper third of the esophagus 4. The thoracic duct 5. The vagi and phrenic nerves and the recurrent laryngeal nerve 6. The great vessels
What are the first branches off of the aorta?	The right and left coronary arteries, which arise just superior to the aortic valve from the right and left aortic sinuses (of Valsalva). There is a third aortic sinus that does not give rise to a vessel (the noncoronary sinus).
What is the first branch off the *arch* of the aorta?	The brachiocephalic artery (called the innominate artery by some clinicians)
The brachiocephalic artery gives off which 2 major branches?	The right common carotid artery and the right subclavian artery
What is the second branch off the arch of the aorta?	The left common carotid artery
What is the third branch off the arch of the aorta?	The left subclavian artery
Which structure hooks around the aortic arch inferior to the ligamentum arteriosum?	The left recurrent laryngeal nerve
The *right* recurrent laryngeal nerve hooks around what structure?	The right subclavian artery. Both recurrent nerves then ascend in the tracheoesophageal groove toward the larynx.

Which 2 veins join to form the superior vena cava?

The left and right brachiocephalic veins

Which 2 veins join to form the brachiocephalic vein?

The internal jugular vein and the subclavian vein

What is the thoracic duct?

A lymphatic channel that conveys lymph from most of the body into the venous system; originates within the abdomen from the chyle cistern

Identify the following structures on this transverse cut just above the aortic arch, as if looking at a computed tomography (CT) or magnetic resonance imaging (MRI) scan (i.e., from the feet up; the left side of the body is the right side of the page):

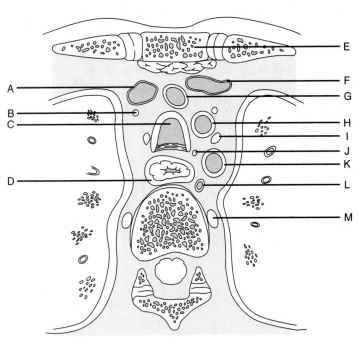

A = Right brachiocephalic vein
B = Phrenic nerve
C = Trachea
D = Esophagus
E = Manubrium
F = Left brachiocephalic vein
G = Brachiocephalic artery
H = Left common carotid artery
I = Vagus nerve
J = Left recurrent laryngeal nerve
K = Left subclavian artery
L = Thoracic duct
M = Sympathetic trunk

Lymph from which part of the body does *not* flow into the thoracic duct?

Right side of head, neck, thorax, mediastinum, and right upper extremity (lymph from these areas conveyed by the right lymphatic duct)

Label the following CT image of the thorax at a similar anatomic level:

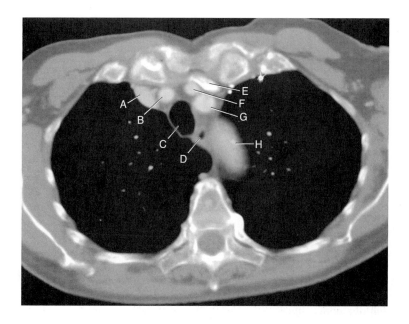

A = Right brachiocephalic ("innominate") vein

B = Brachiocephalic ("innominate") artery

C = Trachea

D = Esophagus

E = Left brachiocephalic ("innominate") vein

F = Left common carotid artery

G = Left subclavian artery

H = Top of aortic arch

The thoracic duct enters the thorax by piercing which part of the diaphragm?

At the aortic hiatus (T12), where it lies just anterior to the vertebral body, between the descending aorta and azygous vein

Describe the course of the thoracic duct within the mediastinum.

Ascends between aorta and azygous vein, crosses from right to left at ~T5, and ascends within superior mediastinum immediately to the left of the esophagus; passes into the neck anterior to the scalenus anterior muscle, posterior to the carotid sheath, and terminates in the neck. Injury to this structure within the thorax can lead to chylothorax (accumulation of lymph fluid or chyle within thoracic cavity).

Where does the thoracic duct empty?

Into the confluence of the left internal jugular and left subclavian veins

Inferior Mediastinum

What are the borders of the anterior mediastinum?

It extends from the sternum to the anterior pericardium, trachea, and great vessels. Note that most clinicians classify the superior mediastinum within the anterior compartment.

What is located in the anterior mediastinum?

A portion of the thymus gland, in addition to lymph nodes and fatty tissue

What are the contents of the middle mediastinum?

1. Pericardium and heart
2. Brachiocephalic vessels
3. Ascending/transverse aorta

4. Superior vena cava (SVC), inferior vena cava (IVC)
5. Pulmonary arteries/veins
6. Trachea/bronchi
7. Nerves, lymph nodes

The inferior mediastinum is divided into which 3 compartments?

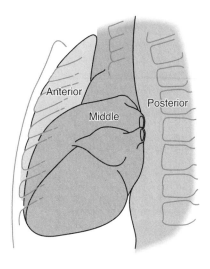

Anterior, middle, and posterior compartments

What is the border between the middle and posterior mediastinum?

The posterior pericardium

List 7 structures contained in the posterior mediastinum.

1. The lower two thirds of the esophagus
2. The anterior and posterior esophageal plexuses
3. The descending aorta
4. The thoracic duct
5. Splanchnic nerves
6. The azygos vein
7. The hemiazygos vein

Which artery supplies the:
   Upper third of the esophagus?

The inferior thyroid artery

Middle third of the esophagus?	The bronchial and esophageal arteries, branches of the descending aorta
Lower third of the esophagus?	The left gastric and left inferior phrenic arteries

Name the most common type of mass found within each mediastinal compartment:

Anterior?	Thymoma
Middle?	Bronchogenic or pericardial cysts
Posterior?	Neurogenic tumors
What are the 5 "birds" of the mediastinum?	1. Vagus nerves ("Va-*goose*") 2. Esophagus ("Esopha-*goose*") 3. Azygos vein ("Azy-*goose*") 4. Hemiazygous ("Hemiazy-*goose*") 5. Thoracic duct ("Thoracic *duck*")

HEART

What are the 3 layers of cardiac tissue?	1. Epicardium 2. Myocardium 3. Endocardium
What is the apex of the heart?	The pointed portion of the heart formed by the projection of the left ventricle inferior, anterior, and to the left
What is the base of the heart?	The posterosuperior portion of the heart formed primarily by the left atrium
What is the margin between the atria and ventricles called?	The coronary sulcus (aka atrioventricular [AV] groove)
What is the margin between the 2 ventricles called?	The interventricular groove

Which cardiac chamber lies anterior, directly posterior to the sternum?

The right ventricle. This chamber's location makes it the most common site for cardiac injury from trauma.

What are the 2 types of pericardium that surround the heart?

1. Fibrous pericardium (the tough outer coat)
2. Serous pericardium (consisting of the visceral serous pericardium, on the surface of the heart itself, and the parietal serous pericardium, on the interior of the fibrous pericardium)

Where is the pericardial fluid located?

In the pericardial cavity, a potential space between the visceral serous pericardium and the parietal serous pericardium

What is pericardial tamponade?

Increases in pericardial fluid (most often blood from traumatic injury) can prevent adequate filling and contraction of the heart, leading to heart failure and death without urgent treatment.

CHAMBERS

What are the muscular ridges on the surfaces of the ventricular walls called?

Trabeculae carneae

What are the 2 portions of the ventricular septum?

The muscular portion and the membranous portion. Ventricular septal defect (VSD) is the most common congenital abnormality of the heart, and most often involves the membranous portion of the septum. VSD leads to abnormal flow from the left ventricle to the right ventricle, and may lead to heart failure if large.

What is the trabeculated muscle on the internal surface of the right atrium called?

Pectinate muscle

What is the smooth posterior wall of the right atrium called?	The sinus venarum
What is the small depression in the interatrial septum (on the right atrial side) called?	The fossa ovalis
The fossa ovalis is a remnant of which embryologic structure?	The foramen ovale. If this structure remains patent into adulthood, it forms the most common type of atrial septal defect (ASD), resulting in the mixing of blood between the left and right atria.

VALVES

Which 2 valves are the:	
Inflow valves of the ventricles?	1. The tricuspid (right AV) valve 2. The mitral (left AV) valve
Outflow valves of the ventricles?	1. The pulmonic valve (right ventricle) 2. The aortic valve (left ventricle)
What are the sources of blood flow into the right atrium of the heart?	The SVC, IVC, and coronary sinus
Describe the flow of blood through the heart.	Blood flows from the right atrium to the right ventricle through the tricuspid valve. From the right ventricle, blood flows into the pulmonary artery through the pulmonic valve. After passing through the capillary beds of the lungs, blood returns to the left atrium via the pulmonary veins. From the left atrium, blood passes to the left ventricle through the mitral valve, and then into the aorta through the aortic valve.
What are the valves formed from?	Thin sheets of fibrous tissue called cusps (leaflets)
How many leaflets does the mitral valve have?	2 (the only valve with 2 cusps!)

What is the orientation of the 2 leaflets of the mitral valve?

Anterior and posterior

 What is rheumatic fever?

A systemic inflammatory disease resulting from infection with Group A *Streptococcus* (strep throat!). This disease affects the heart valves and was extremely common before modern antibiotics. It is still the most common indication for mitral valve repair or replacement.

How many leaflets does the tricuspid valve have?

3

What is the orientation of the 3 leaflets of the tricuspid valve?

Anterior, posterior, and septal (medial)

Each mitral and tricuspid valve leaflet is attached to how many papillary muscles?

2

What structures connect the mitral and tricuspid valve leaflets to the papillary muscles?

Chordae tendineae

What is the function of the papillary muscles?

They help hold the ventricular inflow valves closed during ventricular ejection (contraction). Rupture of these muscles after myocardial infarction ("heart attack") may cause acute valvular insufficiency

Does the pulmonic valve lie anterior or posterior to the aortic valve?

Anterior to it

How many leaflets does the pulmonic valve have?

3

What is the orientation of the 3 leaflets of the pulmonic valve?	Right, left, and anterior
How many leaflets does the aortic valve have?	3
What is the orientation of the 3 leaflets of the aortic valve?	Right, left, and posterior
What is aortic stenosis?	Acquired or congenital obstruction of blood flow across the aortic valve; may require valve replacement in symptomatic or severe cases

CONDUCTING SYSTEM

Where is the sinoatrial (SA) node located?	At the junction of the SVC and the right atrium
What constitutes the blood supply of the SA node?	The SA nodal artery, which in 60% of patients is a branch of the right coronary artery
Failure of the SA node to create electrical impulses may lead to what condition?	Heart block, which may require the implantation of a synthetic pacemaker in severe cases
Where is the AV node located?	In the atrial septum, near the opening of the coronary sinus
What constitutes the blood supply of the AV node?	The AV node is usually supplied by the AV nodal artery, a branch of the right coronary artery.
What are the modified myocardial cells that are responsible for the rapid conduction of electrical impulses through the heart called?	Purkinje fibers

Identify the structures that form the heart's conducting system on the following figure:

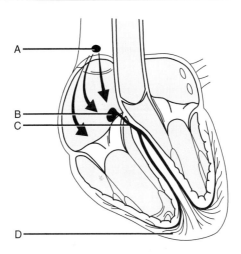

A = Sinoatrial node
B = Atrioventricular node
C = Atrioventricular bundle (of His)
D = Purkinje fibers

What is the path of electrical stimulation in the normally functioning heart?

The impulse arises spontaneously in the **SA node** and travels across the atria, causing them to contract. The impulse stimulates the **AV node**, which distributes the impulse through the **AV bundle (of His)**. The AV bundle (of His) has right and left branches consisting of collections of specialized conducting myocytes called **Purkinje fibers**. From the Purkinje fibers, the impulse passes to the **ventricular myocytes**, resulting in ventricular contraction.

What is atrial fibrillation?

A common clinical condition where abnormal electrical pathways in the heart cause rapid, disorganized atrial contraction, with inconsistent transmission of impulses to the ventricles and an "irregularly irregular" heartbeat

VASCULATURE

Identify the lettered structures corresponding to the arterial supply of the heart on this sternocostal (anterior) view:

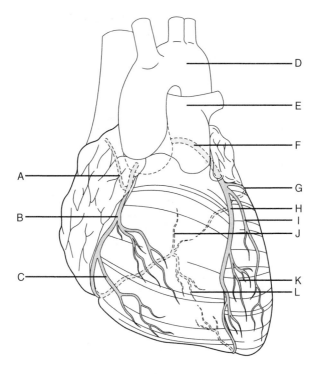

A = Sinoatrial nodal artery
B = Right coronary artery
C = Right marginal artery
D = Aortic arch
E = Pulmonary trunk
F = Left (left main) coronary artery
G = Circumflex artery

H = Anterior interventricular (left anterior descending) artery
I = Left marginal artery
J = Atrioventricular nodal artery
K = Diagonal artery
L = Posterior interventricular (posterior descending) artery

Where do the right and left coronary arteries arise?

From the sinuses of Valsalva, which are located within the right and left leaflets of the aortic valve

The right coronary artery generally gives rise to which 2 major branches?

1. The (acute) marginal artery
2. The posterior interventricular (posterior descending) artery

The anterior right atrial branch of the right coronary artery gives rise to which small, yet very important, branch?

The SA nodal artery

The right coronary artery runs within what groove of the heart?

The coronary sulcus (AV groove)

The left main coronary artery gives rise to which 2 major branches?

1. The circumflex artery
2. The anterior interventricular (left anterior descending) artery

What are the 2 major types of branches of the left anterior descending (LAD) artery?

The diagonal branches and the perforating (septal) branches

Which structure is supplied by the diagonal branches of the LAD?

The anterolateral wall of the left ventricle

Which structure is supplied by the septal branches of the LAD?

The ventricular septum

The LAD artery travels within which groove?

The interventricular groove. It generally anastomoses (communicates) with the posterior interventricular branch of the right coronary artery within this groove.

What are the major branches of the left circumflex coronary artery?

Obtuse marginal branches, which supply the lateral aspect of the left ventricle

The circumflex coronary artery runs within which groove?

The coronary sulcus (AV groove)

 What is a CABG?

Coronary Artery Bypass Grafting; surgical bypass of blocked coronary arteries

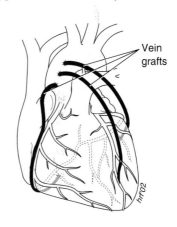

Vein grafts

All of the major cardiac veins terminate in which structure?

The coronary sinus, located in the posterior coronary sulcus (drains into right atrium)

Which cardiac vein travels with the:

    Posterior descending artery?

The middle cardiac vein

    Right marginal artery?

The small cardiac vein

Which vein travels in the anterior interventricular sulcus with the LAD artery?

The great cardiac vein

What are Thebesian veins?

Small veins throughout the heart wall that drain independently into the heart chambers (aka venae cordae minimi)

Identify the shaded structures corresponding to the venous drainage of the heart, as well as the other labeled structures, on the following sternocostal view of the heart:

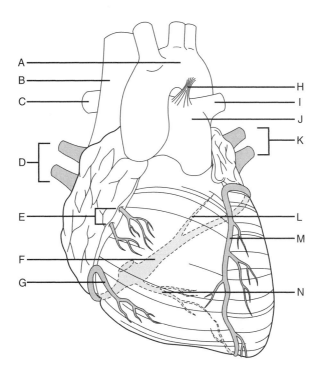

A = Aortic arch
B = Superior vena cava
C = Right pulmonary artery
D = Right pulmonary veins
E = Anterior cardiac veins
F = Coronary sinus
G = Small cardiac vein
H = Ligamentum arteriosum
I = Left pulmonary artery
J = Pulmonary trunk
K = Left pulmonary veins
L = Oblique cardiac vein
M = Great cardiac vein
N = Middle cardiac vein

Where is the coronary vein located?	The coronary vein is another name for the left gastric vein, which takes a circular course along the lesser curvature of the stomach to anastomose with the right gastric vein, forming a ring (corona = crown or ring). Both the left and right gastric veins drain into the portal vein.

INNERVATION

What is the efferent innervation of the heart?	The vagus nerve provides parasympathetic innervation to the SA and AV nodes and the coronary arteries. Sympathetic innervation to these structures is provided by cardiac nerves.
What is the effect of vagal stimulation of the heart?	A decrease in heart rate
What is the afferent innervation of the heart?	The vagus nerve carries afferent impulses to the cardiovascular reflex centers. Pain (such as that from ischemia or infarction) travels with the sympathetic fibers.
Describe the location and contents of the cardiac plexus.	Located anterior to the tracheal bifurcation and posterior to the aortic arch; contains branches of vagus nerve and cervical sympathetic trunk
What is the function of the nerves of the cardiac plexus?	Regulate rate and force of heart muscle contraction and mediate the heart's vascular smooth muscle tone

LUNGS

Name the 2 fissures of the right lung.	Oblique (major) and horizontal (minor)
What is the fissure of the left lung?	The oblique (major) fissure
Name the 3 lobes of the right lung.	1. Right upper lobe 2. Right middle lobe 3. Right lower lobe

Name the 2 lobes of the left lung.

1. Left upper lobe
2. Left lower lobe

What is the lingula?

The lingula (from Latin lingua, meaning "tongue") is a small extension of the left upper lobe. You can think of it as being the left-sided equivalent of a middle lobe.

Name 6 components of the root of the lung.

1. Primary (main stem) bronchus
2. Pulmonary artery
3. Superior and inferior pulmonary veins
4. Bronchial artery and vein
5. The pulmonary nerve plexus
6. Lymphatics

Which of these structures is posterior?

The primary (main stem) bronchus

Which of these structures is anterior and inferior?

The pulmonary veins

How does the location of the pulmonary artery differ from the right to the left?

On the left, the pulmonary artery is superior to the primary (main stem) bronchus (posterior) and the pulmonary vein (anterior). On the right, the pulmonary artery is between the primary (main stem) bronchus (posterior) and the pulmonary vein (anterior).

Where is the vagus nerve located relative to the root of the lung?

Posterior

How about the phrenic nerve?

Anterior

Describe the 10 terms used to describe the airways as they get progressively smaller.

1. Trachea
2. Primary (main stem) bronchus
3. Secondary (lobar) bronchus
4. Tertiary (segmental) bronchus
5. Bronchiole
6. Terminal bronchiole
7. Respiratory bronchiole
8. Alveolar duct
9. Alveolar sac
10. Alveolus

Note that the terminal bronchiole is not, in fact, the end of the line!

PLEURAE

What are the 2 major pleurae of the lungs?

The visceral pleura (which invests the lungs) and the parietal pleura (which lines the thoracic wall)

What is the pleural cavity?

A potential space between the parietal and visceral pleura that contains a small amount of pleural fluid. A pleural effusion is the accumulation of an abnormal amount of pleural fluid.

 **What is a pneumothorax?**

Air within the pleural cavity (normally only a potential space), resulting from an injury to the lung or chest wall. If a large amount of air is trapped within the pleural space, it can compress the vena cavae and impair blood return to the heart, a life-threatening condition termed tension pneumothorax.

Name the 4 types of parietal pleura (as designated by location).

1. Mediastinal pleura
2. Costal pleura
3. Diaphragmatic pleura
4. Cervical pleura (pleural cupula)

The sleeve of pleura just inferior to the root of the lung is called what?

The inferior pulmonary ligament; attaches lower lobe of lung to the mediastinum

Which nerves innervate the parietal pleura?

The intercostal nerves innervate the costal and peripheral diaphragmatic pleurae, and the phrenic nerves innervate the mediastinal and central diaphragmatic pleurae.

The visceral pleura receives somatic afferents from which nerve?

None! There is no somatic sensation of the visceral pleura. Instead, autonomic innervation sensitizes the visceral pleura to stretch.

VASCULATURE

The bronchi and pulmonary connective tissue receive oxygenated blood from which arteries?	The bronchial arteries, branches of the thoracic aorta
What is the venous drainage of the lung?	The 4 pulmonary veins (carrying oxygenated blood) drain into the left atrium. In addition, there are left and right bronchial veins, which drain deoxygenated blood from the tissue near the hilum of the lung into the azygos and hemiazygos veins.
Describe the 3 main components of the azygos venous system.	1. The **azygos vein** proper forms a bridge between the SVC and the IVC and accepts flow from the intercostal veins. Like the vena cavae, the azygos vein runs on the right side. 2. The **hemiazygos vein** drains the inferior left side and empties into the azygos vein. 3. The **accessory hemiazygos vein** drains the superior left side before emptying into the azygos vein. (Azygous is Latin for "unpaired.")
Describe the course of the azygous vein.	Originates in abdomen, posterior to the IVC; enters thorax through aortic hiatus of diaphragm (T12), ascends posterior to aorta and thoracic duct, and empties into the SVC. Note that the azygous runs on the right side of vertebral column, and hemiazygous on the left.
What is the name of the lymph nodes found: **Within the pulmonary parenchyma?**	Pulmonary nodes
At the hilum?	Bronchopulmonary nodes
At the carina?	Tracheobronchial nodes

Tracheobronchial nodes on the right side drain to which structure?

The right lymphatic duct

On the left side?

The thoracic duct

INNERVATION

Describe the location and contents of the pulmonary plexus.

Located along pulmonary vessels and primary bronchi in root of lung, continuous with the cardiac plexus; contains pulmonary branches of vagus nerves and sympathetic trunks

What is the function of the nerves of the pulmonary plexus?

Regulate muscle and glands of bronchial tree and vascular smooth muscle of lungs

Constriction of bronchial muscle leads to what clinical signs and symptoms?

Bronchoconstriction may cause dyspnea (difficulty breathing), wheezing, or cough. Attacks of bronchoconstriction occur with asthma; primary treatment includes inhaled bronchodilator and steroid medications.

DIAPHRAGM

What 3 arteries supply blood to the diaphragm?

1. Inferior phrenic artery
2. Musculophrenic artery
3. Pericardiacophrenic artery

Which nerve innervates the diaphragm?

The phrenic nerve; derived from spinal levels C3–C5, explaining why spinal cord injuries superior to this level may be fatal

Describe the course of the phrenic nerve.

Travels in neck along lateral border of scalenus anterior; enters thorax deep to subclavian vein; runs along the external surface of the fibrous pericardium anterior to the root of the lung, en route to the diaphragm. Because this nerve runs vertically along the pericardium, the pericardium must be opened vertically when accessing the heart during a thoracotomy to avoid injury and subsequent diaphragmatic paralysis.

Which vessels does the phrenic nerve travel with through the thorax?

The pericardiophrenic artery and vein, small branches of the internal thoracic artery

What are the 3 openings in the diaphragm that allow structures to pass from the thoracic cavity into the abdomen?

1. Aortic hiatus
2. Esophageal hiatus
3. Caval foramen

Which 3 structures pass through the aortic hiatus?

1. Aorta
2. Thoracic duct
3. Azygos vein
Think Red, White, and Blue!!

The aortic hiatus is located at which vertebral level?

T12

Which structures form the aortic hiatus?

The right and left crura and the median arcuate ligament

Aside from the esophagus, name 3 structures that pass through the esophageal hiatus.

1. Vagal trunks
2. Esophageal branches of the left gastric vessels
3. Lymphatics from the inferior third of the esophagus

The esophageal hiatus is located at which vertebral level?

T10

 What is a hiatal hernia?

Congenital abnormal enlargement of the esophageal hiatus; may be associated with impaired function of the lower esophageal sphincter and abnormal reflux of gastric acid into the distal esophagus (gastroesophageal reflux disease [GERD])

Which structures pass through the caval foramen?	1. IVC 2. Terminal branches of the right phrenic nerve
The caval foramen is located at which vertebral level?	T8
Diaphragmatic irritation (e.g., from subphrenic abscess) leads to the sensation of pain in what location?	The shoulder! Based on common embryologic derivation

BREAST

In men, the nipple typically lies in which intercostal space?	The fourth
In women, which structures does milk drain through to exit the breast?	Each lobe drains into a lactiferous duct, which in turn drains into a lactiferous sinus. Milk drains from the lactiferous sinus through the nipple to the outside of the body.
What are the ligaments within the breast called?	Cooper's ligaments, or the suspensory ligaments of the breast
What are the 3 major arteries that supply blood to the breast?	1. Perforating branches of the internal mammary (internal thoracic) artery 2. Lateral mammary branches of the lateral thoracic artery 3. Pectoral branches of the thoracoacromial artery
What percentage of breast lymph node drainage is lateral?	Approximately 75%
Where do the lateral lymph nodes of the breast drain to?	Most drain to the axillary lymph nodes, first to the pectoral group (deep to the pectoralis major muscle along the inferior border of the pectoralis minor muscle), and then to the superficial apical

group. Surgery for invasive breast cancer involves removal of axillary lymph nodes in addition to breast parenchyma.

Where do the medial lymph nodes of the breast drain to?

The parasternal lymph nodes, which run with the internal thoracic vein

What is the name of the lymph nodes located between the pectoralis major and pectoralis minor muscles?

Rotter's nodes

POWER REVIEW

THORAX

How do the neurovascular bundles run along the thoracic cage?	In the costal grooves inferior to the ribs
What are the first branches of the aorta?	The right and left coronary arteries
What are the branches of the aortic arch?	The brachiocephalic artery, the left common carotid artery, and the left subclavian artery
Which are the most anterior of the great vessels?	The brachiocephalic veins, which merge to form the SVC
Which structure hooks around the aortic arch just inferior to the ligamentum arteriosum?	The left recurrent laryngeal nerve
The *right* recurrent laryngeal nerve hooks around what structure?	The right subclavian artery

Where does the thoracic duct:

Originate?	Within abdomen, from chyle cistern
Pierce the diaphragm?	At aortic hiatus (~T12) with azygous vein("red, white, and blue")
Cross from right to left within mediastinum?	T5
Empty?	Into left internal jugular (IJ)/subclavian vein
What are the sources of blood flow into the right atrium of the heart?	The SVC, IVC, and coronary sinus

What are the 2 major arteries of the heart, and which parts of the heart do they supply?	1. The **left coronary artery** gives off 2 branches. The anterior interventricular (LAD) artery supplies the anterior portion of the left ventricle and the anterior two thirds of the interventricular septum, and the left circumflex coronary artery supplies the lateral wall of the left ventricle. Both branches supply the left atrium. 2. The **right coronary artery** supplies the right atrium, interatrial septum, right ventricle, posterior wall of the left ventricle, and the posterior third of the interventricular septum.
Which artery supplies the SA node?	The SA nodal artery, usually a branch of the right coronary artery
Which heart valve has 2 cusps?	The mitral valve
What forms the base of the heart?	The left atrium
What are the fissures of the left and right lungs?	**Left:** Oblique (major) fissure (divides the lung into 2 lobes) **Right:** Horizontal (minor) and oblique (major) fissures (divide the lung into 3 lobes)
What is the reflection of pleura inferior to the root of the lung called?	The pulmonary ligament
Which is posterior in the root of the lung, the primary (main stem) bronchus or the pulmonary artery?	The primary (main stem) bronchus
What are the only arteries in the body to carry deoxygenated blood?	The pulmonary arteries

Which vessels form the dual blood supply to the lungs?

The pulmonary arteries (which carry deoxygenated blood) and the bronchial arteries (direct branches of the aorta, which carry oxygenated blood)

At which levels do the 3 major structures pass through the diaphragm?

1. Inferior vena cava: T8
2. Esophagus: T10
3. Aorta: T12

Think **"A, E, I"—alphabetically up the spine**)

Innervation of the diaphragm is by which spinal nerves?

C3, C4, and C5 "keep the diaphragm alive," via the phrenic nerve.

---

9 The Abdomen

Identify the bony landmarks of the abdomen on the following figure:

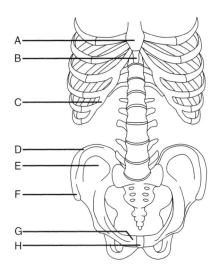

A = Body of sternum   E = Iliac fossa
B = Xiphoid process   F = Anterior superior iliac spine
C = Costal cartilages   G = Pubic tubercle
D = Iliac crest   H = Pubic symphysis

Identify the 9 abdominal regions on the following figure:

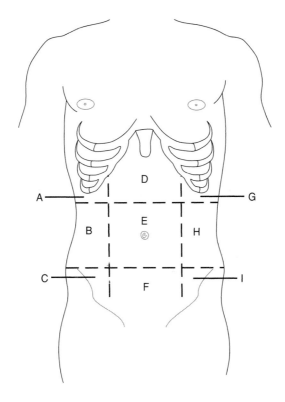

A = Right hypochondriac region
B = Right lumbar region
C = Right inguinal region
D = Epigastric region
E = Umbilical region
F = Suprapubic region
G = Left hypochondriac region
H = Left lumbar region
I = Left inguinal region

Note that in practice, clinicians more often refer to the abdomen in terms of 4 quadrants, delineated by horizontal and vertical planes through the umbilicus.

ABDOMINAL WALL

ANTERIOR ABDOMINAL WALL

Name the 9 layers of the anterior abdominal wall, from superficial to deep.

1. Skin
2. Camper's fascia (fatty layer of the superficial fascia)
3. Scarpa's fascia (membranous layer of the superficial fascia)
4. External oblique muscle
5. Internal oblique muscle
6. Transversus abdominis muscle
7. Transversalis fascia
8. Preperitoneal fatty tissue
9. Peritoneum

What represents the continuation of Scarpa's fascia in the:

Thigh?

The fascia lata

Perineum?

Colles' fascia

Name the 3 flat muscles of the anterolateral abdominal wall.

1. External oblique muscle
2. Internal oblique muscle
3. Transversus abdominis muscle

Describe the direction that the fibers course in the:

External oblique muscle

Inferomedially ("*hands in pockets* because it's cold *outside*)

Internal oblique muscle

Inferolaterally (at right angles to the fibers of the external oblique muscle)

Transversus abdominis muscle

Transversely (horizontally)

Where do the internal oblique and transversus abdominis muscles insert?

In textbooks, the internal oblique and transversus abdominis muscles insert into the pubic tubercle via the conjoint tendon.

In actuality, this structure rarely exists; more commonly, the transversus abdominis aponeurosis alone attaches to the pubic crest.

Name the straplike vertical muscle of the anterolateral abdominal wall.

The rectus abdominis muscle

What structure encloses most of the rectus abdominis muscle?

The rectus sheath

What forms the rectus sheath?

The aponeuroses of the external oblique, internal oblique, and transversus abdominis muscles

In addition to the rectus abdominis muscle, what structures are enclosed by the rectus sheath?

1. The pyramidalis muscle (in 80% of people)
2. The superior and inferior epigastric arteries and veins
3. Lymphatic vessels
4. The T7–T12 ventral primary rami

Which vessel gives rise to the:

    Superior epigastric artery?

Internal thoracic artery

    Inferior epigastric artery?

External iliac artery

What is the significance of the inferior epigastric artery with respect to inguinal hernias?

Direct inguinal hernias occur *medial* to this vessel. *Indirect* hernias are *lateral* to it.

Which approximate dermatomal level is represented by the:

    Xiphoid process?

T7

    Umbilicus?

T10

    Inguinal ligament?

L1

Identify the lettered structures on the following figure of the anterior abdominal wall:

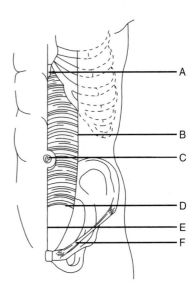

A = Xiphoid process
B = Linea semilunaris
C = Umbilicus
D = Arcuate line
E = Linea alba
F = Inguinal ligament

What is the linea alba?

The midline fascial band that extends from the symphysis pubis to the umbilicus. This is the site of the most common abdominal surgical incision (midline laparotomy).

What is the linea semilunaris?

The curved groove just lateral to the rectus abdominis muscle

What is the arcuate line?

A horizontal line midway between the symphysis pubis and the umbilicus that delineates the transition from the aponeurotic posterior wall of the rectus sheath to the transversalis fascia

How does the rectus sheath differ:

Above the arcuate line?

The aponeurosis of the external oblique muscle contributes to the anterior layer of the sheath, the aponeurosis of the transversus abdominis muscle contributes to the posterior layer, and the aponeurosis of the internal oblique muscle contributes to both layers.

Below the arcuate line?

The aponeuroses of all 3 flat muscles form the anterior layer of the rectus sheath, and the rectus abdominis muscle lies in direct contact with the transversalis fascia posteriorly.

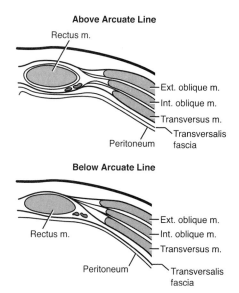

The intercostal nerves run between which layers of the abdominal wall?

The intercostal nerves run between the internal oblique and transversus abdominis muscles, in the so-called "neurovascular plane."

Identify the borders of the inguinal triangle (Hesselbach's triangle) on the following figure:

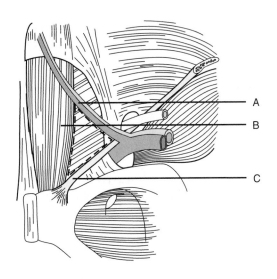

A = Inferior epigastric artery and vein
B = Lateral border of the rectus abdominis muscle (i.e., the linea semilunaris)
C = Inguinal ligament

What is the clinical significance of this triangle?

It marks the site of a *direct* inguinal hernia, an acquired defect in the transversalis fascia of the abdominal wall (in contrast to indirect hernias, which are congenital and pass through the internal inguinal ring, lateral to the inferior epigastric vessels).

What are the major nerves of the anterior abdominal wall?

The inferior 6 thoracic nerves (T7–T11) and the subcostal nerve (T12)

Which muscles are innervated by the subcostal nerve?

The external oblique, internal oblique, transversus abdominis, rectus abdominis, and pyramidalis muscles

Describe the lymphatic drainage of the anterior abdominal wall.

Above the umbilicus, drainage is to the axillary nodes. Below the umbilicus, drainage is to the superficial inguinal, external iliac, and aortic (lumbar) nodes.

POSTERIOR ABDOMINAL WALL

Identify the anterolateral and posterior abdominal wall muscles on the following figure:

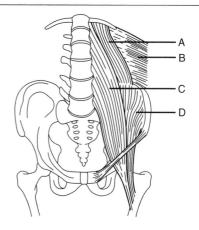

A = Quadratus lumborum muscle
B = Transversus abdominis muscle
C = Psoas major muscle
D = Iliacus muscle

What is the origin and insertion of the psoas major muscle?

The psoas major muscle runs from the transverse processes of the lumbar vertebrae and the bodies of vertebrae T12–L5 to the lesser trochanter of the femur.

 What is the psoas sign?

Because the appendix often lies in direct contact with the psoas muscle, patients with appendicitis may experience pain with extension of the right hip joint, which stretches the iliopsoas muscle.

Where does the iliacus muscle insert?

On the lesser trochanter of the femur (along with the psoas major muscle)

The medial arcuate ligament of the diaphragm is formed by a thickening of which fascia superiorly?

The iliac fascia

The lateral arcuate ligament of the diaphragm is formed by a thickening of which fascia superiorly?

The fascia of the quadratus lumborum muscle

The lumbar plexus lies within which muscle?

The psoas major muscle

What nerves contribute to the lumbar plexus?

The L1–L3 ventral primary rami and the superior branch of spinal nerve L4

What are the 2 largest branches of the lumbar plexus?

The femoral nerve and the obturator nerve

What are the 4 other branches of the lumbar plexus?

1. Ilioinguinal nerve
2. Iliohypogastric nerve
3. Genitofemoral nerve
4. Lateral femoral cutaneous nerve

Describe the course of the femoral nerve.

The femoral nerve arises at the lateral border of the psoas major muscle, descends in the groove between the psoas major and iliacus muscles, and enters the femoral triangle deep to the inguinal ligament (lateral to the femoral vessels).

Which muscles are supplied by the femoral nerve?

The knee extensors

Describe the course of the obturator nerve.

The obturator nerve descends through the psoas major muscle and pierces the fascia to pass lateral to the internal iliac vessels and ureter. The obturator nerve leaves the pelvis and enters the lateral thigh via the obturator foramen.

Which muscles are supplied by the obturator nerve?

The thigh adductors

What is an obturator hernia?

Herniation of bowel through the obturator foramen (right > left), which is normally closed by fat and connective tissue; often leads to bowel obstruction; very difficult clinical diagnosis

 What is the Howship-Romberg sign?

Irritation of the obturator nerve with an obturator hernia may cause pain along the medial thigh with thigh abduction, extension, or internal rotation

Which 2 nerves arise from the L1 segment of the lumbar plexus?

The ilioinguinal and iliohypogastric nerves

Identify the lettered structures on the following illustration of the posterior abdominal wall:

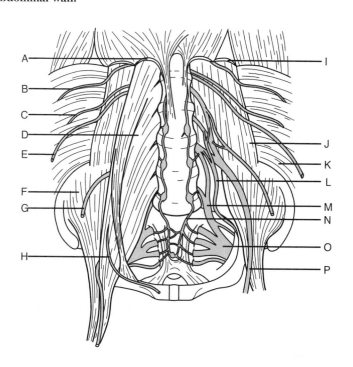

A = Medial arcuate ligament
B = Subcostal nerve
C = Iliohypogastric nerve
D = Psoas major muscle
E = Ilioinguinal nerve
F = Iliacus muscle
G = Lateral femoral cutaneous nerve
H = Genitofemoral nerve
I = Lateral arcuate ligament
J = Quadratus lumborum muscle
K = Transversus abdominis muscle
L = Obturator nerve
M = Lumbosacral trunk
N = Sympathetic trunk
O = Sciatic nerve
P = Femoral nerve

Which structures are innervated by the:

Ilioinguinal nerve?

The abdominal wall muscles and the skin of the groin and genitalia. This nerve is commonly injured or divided during inguinal hernia repair, with a resultant sensory deficit.

Iliohypogastric nerve?

The abdominal wall muscles and the skin of the hypogastric and gluteal regions

Which nerve from the lumbar plexus accompanies the spermatic cord (in men) or the round ligament of the uterus (in women) through the superficial inguinal ring?

The ilioinguinal nerve

Which nerve arises from the L1 and L2 segments of the lumbar plexus?

The genitofemoral nerve

INGUINAL REGION

The inguinal ligament (Poupart's ligament) extends between which 2 structures?

The anterior superior iliac spine and the pubic tubercle

What forms the inguinal ligament?

The folded lower margin of the aponeurosis of the external oblique muscle

Which structures form the:
 Superficial (external) inguinal ring?

The aponeurosis of the external oblique muscle

 Deep (internal) inguinal ring?

The transversalis fascia. The internal ring is the site of an "indirect" inguinal hernia.

What forms the walls of the inguinal canal:
 Anteriorly?

The aponeuroses of the external and internal oblique muscles

 Posteriorly?

The transversalis fascia

 Superiorly?

The arching fibers of the internal oblique and transversus abdominis muscles

 Inferiorly?

The inguinal ligament

Which major structures pass through the inguinal canal in:
 Men?

The spermatic cord and the ilioinguinal nerve

 Women?

The round ligament of the uterus and the ilioinguinal nerve

What structures comprise the spermatic cord?

1. The ductus deferens (vas deferens)
2. The arteries of the ductus deferens (2)
3. The testicular artery
4. The testicular vein (continuous with the pampiniform plexus)
5. The genital branch of the genitofemoral nerve
6. Autonomic nerves
7. Lymphatic vessels draining the testes

 What is a varicocele?

Dilation of veins in spermatic cord (no valves) and pampiniform plexus; may be associated with scrotal pain or with infertility

Which 3 structures form the covering of the spermatic cord?

1. The internal spermatic fascia
2. The cremaster muscle and cremasteric fascia
3. The external spermatic fascia

From what abdominal muscle layer is the cremaster muscle derived?

Internal oblique muscle

From which layer is the inguinal ligament derived?

External oblique muscle

What structure normally attaches the testicle to the scrotum?

The gubernaculum. Testicular torsion is twisting of the testicle around its vascular supply, associated with incomplete attachment of the gubernaculum. It is treated with emergent "detorsion."

What is the orientation of the femoral vessels and nerve inferior to the inguinal ligament?

From lateral to medial: **NAVEL**
Nerve
Artery
Vein
Empty space
Lymphatics

 What is a femoral hernia?

Herniation of abdominal contents inferior to the inguinal ligament (through the "empty space")

PERITONEUM

What is the difference between parietal and visceral peritoneum?

The parietal peritoneum lines the inner abdominal wall and diaphragm. The visceral peritoneum covers the intraabdominal organs. Irritation of the parietal peritoneum by inflammation (e.g., appendicitis) causes "somatic" pain, which is usually well localized. In contrast, "visceral" pain resulting from distension of a hollow viscous (e.g., bowel obstruction) and its visceral peritoneum tends to be vague in character/location.

What is the peritoneal cavity?

The potential space between the 2 layers of peritoneum. This potential space normally contains a small amount of fluid (peritoneal fluid), which functions to lubricate the abdominal organs. Ascites is a pathologic increase in the amount of peritoneal fluid (seen in liver disease, etc.).

Is the peritoneal cavity normally a closed space?

Only in men; in women, the peritoneal cavity communicates with the reproductive tract through the infundibulum of the fallopian tubes (this explains how a pelvic infection can ascend into the abdominal cavity in women).

What is a peritoneal ligament?

A doubled fold of peritoneum that extends between 2 organs

What is a peritoneal fold?

Peritoneum that is reflected away from the abdominal wall by underlying blood vessels or obliterated fetal vessels

What is a mesentery?

A doubled fold of peritoneum that connects an abdominal organ to the abdominal wall. Vessels and nerves that supply the organ are housed in between the doubled layers of peritoneum.

What is an "intraperitoneal" organ?

An organ with a mesentery

What is a "retroperitoneal" organ?

An organ that sits directly on the posterior abdominal wall and is covered anteriorly with visceral peritoneum

Which gastrointestinal structures are retroperitoneal?

1. The pancreas
2. Most of the duodenum
3. The ascending and descending colon
Note that the appendix is **not** retroperitoneal in most patients.

Name the mesenteries of the following organs:

Small intestine — Mesentery proper or "mesentery of the small intestine"

Transverse colon — Transverse mesocolon

Sigmoid colon — Sigmoid mesocolon

Appendix — Mesoappendix

Stomach — Greater omentum and lesser omentum

List 6 structures that are crossed by the root of the mesentery proper.

1. Duodenum
2. Aorta
3. Inferior vena cava
4. Psoas major muscle
5. Right ureter
6. Right testicular or ovarian vessels

What are the 5 sites of attachment of the greater omentum?

1. The greater curvature of the stomach
2. The first portion of the duodenum
3. The diaphragm
4. The spleen
5. The transverse colon

What are the 3 sites of attachment of the lesser omentum?

1. The lesser curvature of the stomach
2. The first portion of the duodenum
3. The liver

Which vessels run between the 2 layers of the lesser omentum?

The left and right gastric vessels

What is the name of the space posterior to the stomach and lesser omentum?

The omental bursa, or lesser peritoneal sac

What is the site where the omental bursa opens into the rest of the peritoneal cavity (i.e., the greater peritoneal sac) called?

The epiploic foramen (of Winslow)

What is the name of the right free margin of lesser omentum?

The hepatoduodenal ligament (named aptly for its 2 attachments)

Which 3 important structures lie in the hepatoduodenal ligament?

1. The common hepatic artery
2. The portal vein
3. The common bile duct

The Pringle maneuver is a surgical technique that involves digital compression of these structures, used during liver surgery to reduce bleeding.

What is the orientation of the hepatic artery, portal vein, and bile duct within the hepatoduodenal ligament?

The **P**ortal vein is **P**osterior, the hepatic artery is left anterior, and the bile duct is right anterior. (Think **LARD: L**eft **A**rtery, **R**ight **D**uct.)

Name the double layer of parietal peritoneum that connects the liver to the diaphragm and the anterior abdominal wall.

The falciform ligament

Which blood vessels lie in the falciform ligament?

The paraumbilical veins

Which structure lies in the free margin of the falciform ligament?

The ligamentum teres

What is the embryologic origin of the ligamentum teres?

The fetal left umbilical vein

Which structures form the:
 Median umbilical folds?

A remnant of the urachus

 Medial umbilical folds?

Obliterated fetal umbilical arteries

 Lateral umbilical folds?

Inferior epigastric vessels

What is the most posterior recess within the abdominal cavity?	The hepatorenal recess (aka Morrison's pouch). Because of its location, the hepatorenal recess is a common site for intraabdominal fluid collections and abscesses.

 What is the pouch of Douglas?

Space between the rectum and bladder or uterus; also a common place for abscesses

Where are the:

Paracolic gutters?

The right paravertebral gutter lies on the right side of the ascending colon and forms a communication between the hepatorenal recess and the pelvis. The left gutter lies on the left side of the descending colon and connects the left side of the abdominal cavity with the pelvic cavity.

Paravertebral gutters?

On either side of the vertebral column

ABDOMINAL VASCULATURE

ARTERIES

 What is an abdominal aortic aneurysm?

An abnormal dilation of the abdominal aorta

Which external anatomic landmark lies at the level of the abdominal aortic bifurcation?

The umbilicus. To examine someone for an abdominal aortic aneurysm (AAA), which is usually just above the aortic bifurcation, palpate just above (and to the left of) the umbilicus.

 Which vein often crosses anterior to the neck of an AAA proximally?

The left renal vein

Label the branches of the abdominal aorta on the following figure:

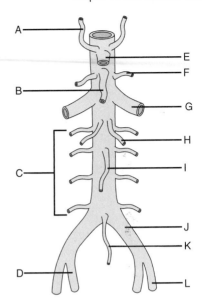

A = Inferior phrenic artery
B = Superior mesenteric artery
C = Lumbar arteries
D = Internal iliac artery
E = Celiac trunk
F = Middle suprarenal artery
G = Renal artery
H = Testicular or ovarian artery
I = Inferior mesenteric artery
J = Common iliac artery
K = Middle (median) sacral artery
L = External iliac artery

What part of the small bowel crosses directly anterior to the aorta at the level of an AAA?

The duodenum. Infected aortic grafts may erode into the duodenum, leading to gastrointestinal hemorrhage ("aortoenteric fistula").

Which vein runs posterior to the *right* common iliac artery?

The *left* common iliac vein

Where does the external iliac artery become the femoral artery?

At the inguinal ligament. Remember: The artery runs *lateral* to the femoral vein.

Which 3 arteries arise from the celiac trunk?

1. Splenic artery
2. Left gastric artery
3. Common hepatic artery

Identify the arterial branches of the celiac trunk:

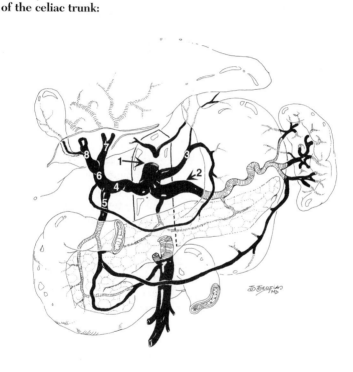

1. Celiac trunk
2. Splenic artery
3. Left gastric artery
4. Common hepatic artery
5. Gastroduodenal artery
6. Proper hepatic artery
7. Left hepatic artery
8. Right hepatic artery

Name the 6 major branches of the superior mesenteric artery.

1. Inferior pancreaticoduodenal artery
2. Middle colic artery
3. Ileocolic artery
4. Right colic artery
5. Jejunal arteries
6. Ileal arteries

Label the branches of the superior mesenteric artery on the following figure:

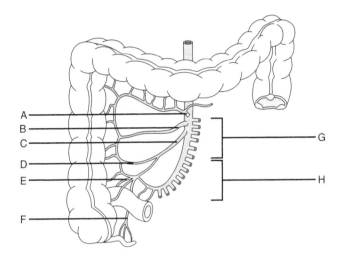

A = Middle colic artery
B = Right colic artery
C = Ileocolic artery
D = Ascending colic branches
E = Cecal branches
F = Appendiceal artery
G = Jejunal arteries
H = Ileal arteries

Name the 4 major branches of the ileocolic artery.

1. Ascending colic branches
2. Cecal branches
3. Appendiceal artery
4. Ileal and jejunal branches

**Label the 3 major branches
of the inferior mesenteric
artery on the following
figure:**

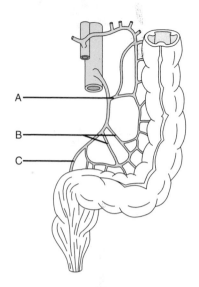

A = Left colic artery
B = Sigmoid arteries (2 or 3 are usually
present)
C = Superior rectal artery

VEINS

**Label the tributaries of the
abdominal inferior vena cava
on the following figure:**

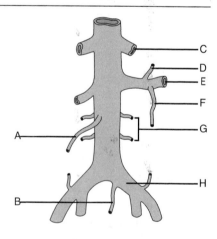

A = Right testicular or ovarian vein
B = Middle sacral vein
C = Hepatic vein
D = Inferior adrenal (suprarenal)
vein

E = Left renal vein
F = Left testicular or ovarian vein
G = Lumbar veins
H = Left common iliac vein

What is the drainage of the: **Left testicular or ovarian** **vein?**	Into the left renal vein
Right testicular or **ovarian vein?**	Directly into the inferior vena cava
Which is longer, the left or **right renal vein?**	The left

ABDOMINAL VISCERA

Label the following diagram
of the abdominal organs:

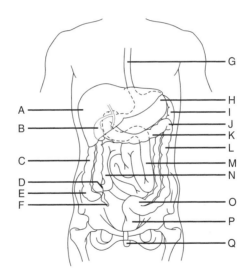

A = Liver
B = Gallbladder
C = Ascending colon
D = Ileocecal valve
E = Cecum
F = Vermiform appendix
G = Esophagus
H = Stomach
I = Spleen

J = Left colic (splenic) flexure
K = Transverse colon
L = Descending colon
M = Jejunum
N = Ileum
O = Sigmoid colon
P = Rectum
Q = Anus

Describe the extent of the:
 Foregut? — From the oropharynx to the hepatopancreatic ampulla (of Vater)

 Midgut? — From the hepatopancreatic ampulla (of Vater) to the distal third of the transverse colon

 Hindgut? — From the distal transverse colon to the anus

What is the major arterial supply of the:
 Foregut? — The celiac trunk

 Midgut? — Superior mesenteric artery

 Hindgut? — Inferior mesenteric artery

ESOPHAGUS

Which structure accompanies the esophagus through the diaphragm? — The vagus nerves (left and right)

Is the left vagus nerve anterior or posterior? — Anterior (Remember LARP: Left Anterior, Right Posterior.)

Which arteries supply the abdominal esophagus? — The inferior phrenic artery and the left gastric artery

What is the venous drainage of the abdominal esophagus? — The azygos vein and the left gastric vein

Name the layers of the esophageal wall (from deep to superficial).

1. Mucosa
2. Submucosa
3. Muscularis

Note that unlike the rest of the upper gastrointestinal tract, the esophagus has no serosa. Because of this, esophageal cancer tends to invade outside structures early in its course, and accordingly is associated with a high mortality rate.

How does the muscular layer of the esophagus differ along its length?

The upper third of the esophagus has striated (voluntary) muscle, the lower third has smooth (involuntary) muscle, and the middle third has both striated and smooth muscle.

Which muscle functions as the upper esophageal sphincter?

The cricopharyngeus muscle

Where are the 4 most likely points of constriction or obstruction along the esophagus?

1. Upper esophageal sphincter
2. Aortic arch
3. Left main stem bronchus
4. Lower esophageal sphincter

The upper esophageal sphincter is the narrowest point in the entire gastrointestinal tract, and is therefore the most common site for a swallowed foreign body to lodge.

The esophagus enters which part of the stomach?

The cardia

STOMACH

What is the anatomic division between the body and antrum of the stomach?

An imaginary line drawn from the incisure/angular notch to the junction of the right and left gastroepiploic arteries. Surgery for peptic ulcer disease may involve an antrectomy, because the antrum is the location of parietal cells, which make hydrochloric acid.

**Identify the labeled
structures on the following
diagram of the stomach:**

A = Esophagus
B = Cardiac notch
C = Lesser curvature of the stomach
D = Angular notch (incisure)
E = Lesser omentum
F = Pylorus
G = Duodenum
H = Fundus
I = Body of the stomach
J = Greater curvature of the stomach
K = Greater omentum
L = Antrum

**What part of the stomach
contacts the diaphragm?**

The fundus

**The anterior stomach
contacts which 3 structures?**

1. The anterior abdominal wall
2. The left lobe of the liver
3. The diaphragm

What is the name of the large, longitudinal folds in the mucous membrane of the stomach?

Rugae. These folds flatten as the stomach fills with food.

What are the 3 muscular layers of the stomach wall, from deep to superficial?

1. Oblique fibers
2. Circular fibers
3. Longitudinal fibers

Which arteries constitute the stomach's arterial supply?

1. The left and right gastric arteries (branches of the celiac artery and the hepatic artery, respectively)
2. The left and right gastroepiploic arteries (branches of the splenic artery and the gastroduodenal artery, respectively)
3. The short gastric arteries (branches of the splenic artery)

Which vessels run along the stomach's:
    Lesser curvature?

The right and left gastric arteries

    Greater curvature?

The right and left gastroepiploic arteries

Identify the labeled arteries on the following figure of the stomach:

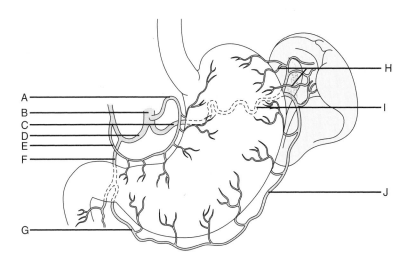

A = Left gastric artery
B = Celiac trunk
C = Splenic artery
D = Hepatic artery
E = Right gastric artery
F = Gastroduodenal artery
G = Right gastroepiploic artery
H = Short gastric arteries
I = Splenic artery
J = Left gastroepiploic artery

Are nerves on the anterior stomach derived from the left or right vagus?

The left (Remember **LARP!**)

Which portion of the stomach lies just proximal to the duodenum?

The pylorus. The muscular pylorus, which regulates the flow of gastric contents into the duodenum, may be overly developed in infancy, requiring surgical division of the hypertrophied muscle (pyloromyotomy).

Which blood vessel marks the gastroduodenal junction and site of the pyloric orifice?

The prepyloric vein (of Mayo)

SMALL INTESTINE

How long is the adult small intestine?

Approximately 20 feet (6 meters)

What are plicae circulares (valvulae conniventes)?

Circular folds of small intestinal mucosa. This feature is useful for differentiating the small bowel from the large bowel on abdominal X-rays (the plicae circulares are seen as complete circles).

Name the 3 parts of the small intestine.

Duodenum, jejunum, and ileum

Duodenum

What is the derivation of the name "duodenum"?

Duodeni is Latin for "twelve"; the duodenum is about 12 fingerbreadths (25 cm) long.

Is the duodenum intraperitoneal or retroperitoneal?

The proximal duodenum (i.e., the first 2.5 cm or so) has a mesentery and is therefore intraperitoneal; the remainder of the duodenum is retroperitoneal.

Which arteries supply the duodenum?

1. The superior pancreaticoduodenal arteries (branches of the gastroduode- nal artery)
2. The inferior pancreaticoduodenal arteries (branches from the superior mesenteric artery)

These vessels anastomose to form arterial arcades.

What are the 4 parts of the duodenum?

1. Superior
2. Descending
3. Inferior (horizontal)
4. Ascending

What is the most common site for peptic ulcer disease (PUD)?

The second portion of the duodenum. Anterior ulcers may present with free perforation of the duodenum (and peritonitis). Posterior ulcers more often lead to gastrointestinal hemorrhage.

What is the approximate vertebral level of the:

    First portion of the duodenum?

Approximately L1

    Third portion of the duodenum?

Approximately L3

What is another name for the beginning of the first portion of the duodenum?

The duodenal bulb (ampulla)

Which artery lies directly behind the first portion of the duodenum?

The gastroduodenal artery (The location of this major artery makes it susceptible to invasion by penetrating posterior ulcers in the first portion of the duodenum, which can lead to gastrointestinal hemorrhage.)

Where do the common bile duct and the pancreatic ducts enter the duodenum?

On the posteromedial aspect of the second part of the duodenum

What is the name of the dilatation formed by the junction of the common bile duct and the main pancreatic duct (of Wirsung) just proximal to their opening into the duodenum?

The hepatopancreatic ampulla (of Vater)

What is the name of the muscular sphincter at the hepatopancreatic ampulla (of Vater)?

The sphincter of Oddi

What does the sphincter of Oddi do?

It regulates the entry of bile and pancreatic enzymes into the small intestine.

What marks the spot on the duodenal wall where the hepatopancreatic ampulla (of Vater) enters the intestinal lumen?

The major duodenal papilla

What is the minor duodenal papilla?

The minor duodenal papilla is a communication between the accessory pancreatic duct (of Santorini) and the duodenum. When present (10% of patients), the minor duodenal papilla usually lies several centimeters above the opening of the main pancreatic duct (of Wirsung). In the rest of the population, the accessory pancreatic duct merges with the main pancreatic duct before it enters duodenum.

Jejunum and Ileum

What is the derivation of the name:	
"Jejunum"?	Latin for "empty"
"Ileum"?	Latin for "twisted"

What is the ligament of Treitz?	A well-marked peritoneal fold at the junction of the fourth part of the duodenum with the jejunum. This marks the site where the small bowel passes through the transverse mesocolon. "Running" the small bowel at surgery involves an examination from the ligament of Treitz to the cecum.

Describe the path of blood flow from the aorta to the jejunum.	The aorta gives rise to the **superior mesenteric artery,** which runs in the root of the mesentery, giving off 15–18 **jejunal** and **ileal branches** that run between the 2 layers of mesentery and unite to form loops called arterial arcades. These arcades then form **straight vessels (vasa recta),** which pass alternately to opposite sides of the intestine and enter the intestine on the mesenteric border.

How does the jejunum differ in appearance from the ileum?	1. Thicker wall
	2. Larger diameter
	3. More vascular (redder)
	4. Larger and more developed plicae circulares
	5. Longer vasa rectae
	6. Less mesenteric fat
	7. Lacks Peyer's patches (lymphoid aggregates along the antimesenteric border of the ileum)

Where are the mesenteric attachments of the:	
Jejunum?	Above and to the left of the aorta

Ileum?

Below and to the right of the aorta
(Accordingly, the jejunum tends to
occupy more of the left upper quadrant
and the ileum tends to occupy more of
the right lower quadrant.)

**Which structure marks the
end of the small intestine?**

The ileocecal valve

LARGE INTESTINE

**How long is the adult large
intestine?**

Approximately 1.5 meters

**Identify the labeled
structures on the following
diagram of the large
intestine:**

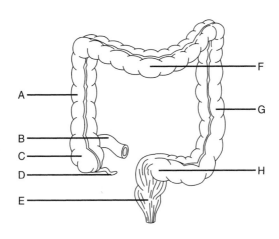

A = Ascending colon
B = Terminal ileum
C = Cecum
D = Vermiform appendix
E = Rectum
F = Transverse colon
G = Descending colon
H = Sigmoid colon

Name the 7 parts of the large intestine, from proximal to distal.

1. Vermiform appendix
2. Cecum
3. Ascending colon
4. Transverse colon
5. Descending colon
6. Sigmoid colon
7. Rectum

Name 3 features that differentiate the large intestine from the small intestine on gross inspection.

1. Teniae coli
2. Haustra
3. Appendices epiploicae

What are the teniae coli?

The 3 thin bands of muscle that run longitudinally along the entire length of the ascending, transverse, and descending colon

Where do the 3 taeniae coli converge?

At the appendix (This can help you locate the appendix during surgery.)

What are haustra, and what causes them?

Characteristic sacculations of the large intestine, which are created by contraction of the taeniae coli. On X-rays, the haustra appear as incomplete circles.

What are appendices epiploica (epiploic appendages)?

The small, fat-filled peritoneal sacs along the taeniae coli

Which parts of the large intestine are retroperitoneal?

The ascending and descending colon

What portion of the large intestine has the widest diameter?

The cecum (typically measures 7–9 cm in diameter). Because of LaPlace's law (wall tension is proportional to radius), the cecum's large diameter makes it the *most* likely site of colonic perforation.

What is the narrowest part of the colon?

The sigmoid. Accordingly, the sigmoid is the most likely site for colonic obstruction.

What arteries supply the cecum?

The anterior and posterior cecal arteries (branches of the ileocolic artery, which is a branch of the superior mesenteric artery)

Where is the vermiform appendix located?

Its position is highly variable. Although it is most commonly retrocecal, the appendix may also be located in the pelvis or retroperitoneum.

What artery supplies the vermiform appendix?

The appendicular artery, a branch of the ileocolic artery

Does the appendix have a mesentery?

Yes, the appendix has a mesentery even though the cecum does not. The mesoappendix suspends the appendix from the mesentery of the terminal ileum.

What is appendicitis?

Inflammation of the appendix caused by obstruction of the appendiceal lumen; the most common reason for emergent abdominal surgery in the United States

What arteries supply the ascending colon?

The ileocolic and right colic arteries (both branches of the superior mesenteric artery)

Where is the point of transition between the superior and inferior mesenteric arterial blood supply to the large intestine?

At the beginning of the distal third of the transverse colon (the site of embryologic transition from midgut to hindgut)

What are the clinical implications associated with this transitional area?

This area, which is between blood supplies, is especially prone to ischemia during times of decreased blood flow to the colon (the so-called "watershed" effect).

What is McBurney's point?

A point two-thirds down along an imaginary line between the umbilicus and the anterior superior iliac spine. This is the classic location of pain and tenderness with appendicitis.

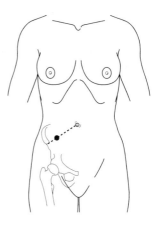

Describe the mesentery of the transverse colon.

The transverse mesocolon is a double layer of peritoneum that suspends the transverse colon from the posterior abdominal wall and connects the transverse colon to the pancreas and greater omentum.

Which arteries supply the descending colon?

The left colic and superior sigmoid arteries (both branches of the inferior mesenteric artery)

What are the white lines of Toldt?

The lateral peritoneal reflections of the ascending and descending colon

How can one grossly identify the beginning of the sigmoid colon?

The descending colon is retroperitoneal and the sigmoid colon is intraperitoneal.

Which arteries supply the sigmoid colon?

The sigmoid arteries

How does the rectum differ from the sigmoid colon?

1. The rectum lacks a peritoneal covering distally.
2. The teniae coli broaden to form a complete longitudinal layer.
3. The rectum has no haustra or appendices epiploicae, but it does have transverse rectal folds (which the sigmoid colon lacks).

PANCREAS

Identify the labeled structures on this figure of the pancreas and surrounding organs:

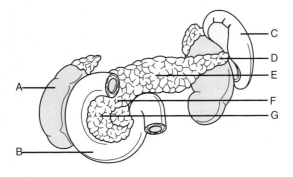

A = Right kidney
B = Duodenum
C = Spleen
D = The tail of the pancreas
E = The body of the pancreas
F = The neck of the pancreas
G = The head of the pancreas

The head of the pancreas lies in intimate contact with which organ? How about the tail?

The head of the pancreas lies in the duodenum, and the tail contacts the spleen (the pancreas is "cradled in the arms of the duodenum and tickles the spleen with its tail").

Why must the duodenum be removed if the head of the pancreas is removed?

They share the same blood supply.

What is the uncinate process?

A hook-shaped projection from the lower aspect of the head of the pancreas that extends superiorly and to the left and lies between the superior mesenteric vessels and the aorta

What landmark defines the neck of the pancreas?

The superior mesenteric vessels. The involvement of these vessels is a major determinant as to whether a pancreatic cancer is "resectable."

Which major vessels course along the pancreas?

The splenic artery (which originates at the celiac trunk) runs along the superior margin of the pancreas, and the splenic vein runs just dorsal to the pancreas. (Inflammation of the pancreas [pancreatitis] can lead to splenic vein thrombosis.)

Which arteries supply the head of the pancreas?

1. The anterior superior and posterior superior pancreaticoduodenal arteries (branches of the gastroduodenal artery)
2. The anterior inferior and posterior inferior pancreaticoduodenal arteries (branches of the superior mesenteric artery)

What arteries supply the body and tail of the artery pancreas?

1. The splenic artery
2. The dorsal pancreatic artery
3. The great pancreatic artery (pancreatica magna)
4. The caudal pancreatic artery

What is the duct of Wirsung? Of Santorini?

The main pancreatic duct and the accessory pancreatic duct, respectively

Identify the labeled arteries on the following figure of the pancreas:

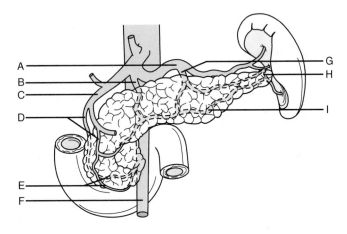

A = Splenic artery
B = Dorsal pancreatic artery
C = Gastroduodenal artery
D = Posterior and anterior superior pancreaticoduodenal arteries
E = Posterior and anterior inferior pancreaticoduodenal arteries
F = Superior mesenteric artery
G = Great pancreatic artery
H = Caudal pancreatic arteries
I = Inferior pancreatic artery

The main pancreatic duct (of Wirsung) normally joins which structure?

The common bile duct

LIVER

Which structure traditionally divides the liver into right and left lobes, anatomically speaking?

The falciform ligament

What does "functionally independent" mean?

The right and left functional lobes have their own arterial and portal blood supply, venous drainage, and biliary drainage.

What line divides the liver into functionally independent right and left lobes?

Cantlie's line (an imaginary line drawn across the diaphragmatic surface of the liver that runs from the gallbladder to the inferior vena cava)

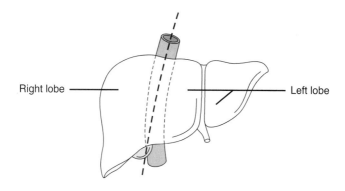

Right lobe — — — Left lobe

What are the 4 lobes of the liver?

1. Right lobe
2. Left lobe
3. Quadrate lobe
4. Caudate

From a functional standpoint, the caudate and quadrate lobes are part of which side of the liver?

The left

What is an alternative scheme for dividing the liver anatomically and functionally?

The French (hepatic segment) system, based on the branching of the portal vein and hepatic artery

Identify the liver segments according to the French system on the figure below:

An easy way to remember the order of the segments: Start by identifying segment 1, then move in a clockwise direction.

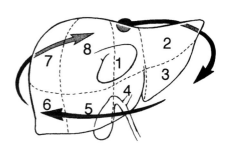

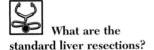 **What are the** See figure
standard liver resections?

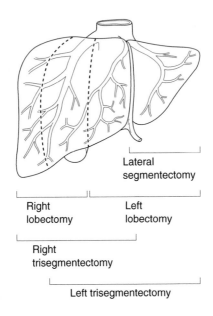

What is the name of the
space between the liver and
the diaphragm?

The subphrenic recess

What is the coronary
ligament?

The peritoneal reflection on the superior
part of the liver that "crowns" the liver
and attaches it to the diaphragm.

What are the triangular
ligaments of the liver?

The right and left lateral extents of the
coronary ligament, which form triangles

What is the "bare area" of
the liver?

The "bare area" of the liver is the
triangular area of the liver that is **not
covered by peritoneum.** The 2 layers
of the falciform ligament separate
superiorly and are reflected onto the
diaphragm as the coronary ligament,
leaving this part of the liver in direct
contact with the diaphragm.

What is the porta hepatis?

A transverse fissure on the visceral surface of the liver; the peritoneum opens at this fissure to transmit the portal vein, hepatic artery proper, and the right and left hepatic bile ducts.

Name the 4 organs that contact the visceral surface of the liver.

1. Stomach
2. Duodenum
3. Gallbladder
4. Colon (at the right colic flexure)

How is the liver attached to the stomach and duodenum?

By the lesser omentum, which has 2 parts, the hepatogastric ligament and the hepatoduodenal ligament

Identify the labeled structures on the following views of the liver and associated area:

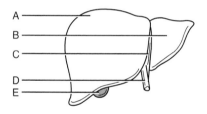

Anterior surface

Posterior surface, liver retracted superiorly

A = Right liver lobe
B = Left liver lobe
C = Falciform ligament
D = Ligamentum teres
E = Gallbladder
F = Gallbladder
G = Quadrate liver lobe
H = Ligamentum teres
I = Porta hepatis
J = Ligamentum venosum
K = Caudate liver lobe
L = Inferior vena cava

Describe the path of arterial blood flow from the aorta to the liver.

The common hepatic artery arises from the celiac trunk and becomes the proper hepatic artery at the takeoff of the gastroduodenal artery. The hepatic artery proper divides into right and left branches shortly before entering the liver at the porta hepatis.

What percentage of blood flow to the liver is provided by the:

Hepatic artery?

Approximately 30%

Portal vein?

The remaining 70%

What percentage of *oxygen* to the liver is provided by the:

Hepatic artery?

Approximately 50%

Portal vein?

The remaining 50%

What is the general function of the portal vein?

The portal vein collects blood from the gastrointestinal tract, gallbladder, pancreas, and spleen, and carries it to the liver.

Which 2 veins unite to form the portal vein?

The superior mesenteric vein and the splenic vein

What happens to the inferior mesenteric vein?

The inferior mesenteric vein most commonly joins the splenic vein before the latter unites with the superior mesenteric vein to form the portal vein.

Name the 4 communications between the portal circulation and the systemic circulation that provide collateral portal circulation in the event of obstruction in the liver or portal vein.

1. The left gastric (coronary) vein anastomoses with the esophageal veins of the azygos system.
2. The superior rectal vein anastomoses with the middle and inferior rectal veins.
3. The paraumbilical veins anastomose with the epigastric veins of the anterior abdominal wall.
4. Tributaries of the splenic and pancreatic veins anastomose with the left renal vein.

In patients with portal hypertension, blood tends to be diverted into the systemic circulation at the sites of the portosystemic anastomoses. What clinical conditions may be seen when this occurs?

Esophageal varices (site #1), hemorrhoids (site #2), and caput medusae (site #3). Think "gut, butt, and caput" plus the retroperitoneum (site #4).

What is the venous drainage of the liver?

The hepatic veins drain directly into the inferior vena cava, just inferior to the diaphragm.

What is the primary lymphatic drainage of the liver?

The lymphatics drain to the hepatic nodes, which are located alongside the hepatic vessels in the lesser omentum. The hepatic nodes then drain to the celiac nodes.

GALLBLADDER AND BILE DUCTS

What is the approximate capacity of the adult gallbladder?

30–50 mL

Label the parts of the gallbladder on the following figure:

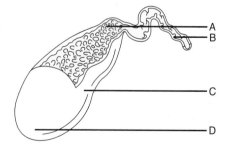

A = Infundibulum of the gallbladder (Hartman's pouch)
B = Cystic duct
C = Body of the gallbladder
D = Fundus of the gallbladder

What are the folds in the mucous membrane of the cystic duct called?

The spiral valves of Heister (impacted gallstones often lodge in the spiral valves)

Label the following figure of the biliary tree:

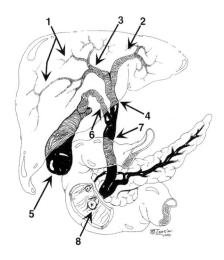

1. Intrahepatic ducts
2. Left hepatic duct
3. Right hepatic duct
4. Common hepatic duct
5. Gallbladder
6. Cystic duct
7. Common bile duct
8. Ampulla of Vater

Which artery supplies the gallbladder?

The cystic artery, which is normally a branch of the right hepatic artery; however, there is tremendous variability in the origin of the cystic artery and other vessels near the porta hepatis.

What is the venous drainage of the gallbladder?

Most of the gallbladder fundus and body drains directly into the liver.

What is the gallbladder's lymphatic drainage?

The lymphatics drain to the hepatic lymph nodes, which in turn drain to the celiac nodes.

Define the cystohepatic (Calot's) triangle.

It's the anatomic triangle bordered by the liver superiorly, the cystic duct inferiorly, and the common hepatic duct medially.

Which key structure lies within Calot's triangle?

The cystic artery. This anatomic relationship is important in locating the cystic artery during laparoscopic cholecystectomy (gallbladder removal).

Trace the flow of bile from the gallbladder to the intestine.

Bile flows out of the gallbladder infundibulum and into the cystic duct. The cystic duct joins the common hepatic duct to form the common bile duct, which travels with the hepatic artery and portal vein in the lower free edge of the lesser omentum. After passing posterior to the superior portion of the duodenum, the common bile duct is usually joined by the main pancreatic duct (of Wirsung); it then enters the hepatopancreatic ampulla (of Vater). The hepatopancreatic ampulla (of Vater) opens into the second portion of the duodenum at the major duodenal papilla, which is about 8–10 cm distal to the pylorus.

What is the narrowest point in the biliary system?

The hepatopancreatic ampulla (of Vater). Gallstones, which are usually formed within the gallbladder, may migrate into the bile ducts and cause obstruction at the ampulla; usually treated endoscopically

Name the 2 muscular sphincters in the distal biliary tract.

1. The choledochal sphincter (located at the distal end of the bile duct)
2. The hepatopancreatic sphincter (of Oddi), located at the hepatopancreatic ampulla (of Vater)

SPLEEN

How big is a normal adult spleen?

About the size of a fist

Which 4 abdominal organs are normally in contact with the spleen?

1. The stomach
2. The left kidney
3. The colon
4. The pancreas

Which of these organs contacts the hilum of the spleen?	The pancreas
What other structure does the spleen lie in direct contact with?	The diaphragm
Which vessels run in the gastrolienal (gastrosplenic) ligament?	The short gastric and the left gastroepiploic vessels
Which vessels run in the lienorenal (splenorenal) ligament?	The splenic artery and vein
Which artery constitutes the spleen's primary blood supply?	The splenic artery, the largest branch from the celiac trunk
Describe the course of the splenic artery.	The splenic artery runs from the celiac trunk along the superior border of the pancreas in the lienorenal ligament. It then divides within the ligament into several branches, which enter the spleen at the hilum.
Describe the spleen's venous drainage.	Venous drainage of the spleen is via the splenic vein, which unites with the superior mesenteric vein posterior to the pancreas to form the portal vein.
What is an accessory spleen?	Heterotopic splenic tissue, usually located near the splenic hilum (seen in approximately 20% of the population)

KIDNEYS, ADRENAL GLANDS, AND URETERS

The kidneys normally lie at approximately what vertebral level?	At the level of the T12–L3 vertebrae
Are both kidneys normally at the same level?	No. Because of the large right lobe of the liver, the right kidney tends to be located more inferiorly than the left.

Are the kidneys retroperitoneal or intraperitoneal?	Retroperitoneal
The kidneys lie in close association with which muscle?	The psoas major muscle
What is the name of the capsule that surrounds the kidney?	The renal fascia (Gerota's fascia)
What lies between the renal fascia and the peritoneum of the posterior abdominal wall?	Pararenal fat
What lies between the renal fascia and the kidney?	Perirenal fat
What is the name of the kidney's medial concave margin?	The renal sinus (the site of the renal hilum and the renal pelvis)
Describe the gross organization of the kidney on cut section.	The renal parenchyma is organized into a cortex and medulla; the cortex is the portion of the parenchyma closest to the renal fascia.
What arteries supply the kidneys?	The renal arteries, which are branches of the abdominal aorta at the level of the L1–L2 vertebrae
Name the 3 structures that enter or leave the kidney at its hilum.	1. Renal vein 2. Renal artery 3. Renal pelvis
What is the orientation of these structures?	The renal vein is anterior, the renal artery is posterior, and the renal pelvis is posterior to the artery. ("VAP-AP": **V**ein-**A**rtery-**P**elvis, **A**nterior to **P**osterior)
What is the renal pelvis?	The area of transition between the kidney and the ureter
Trace the flow of urine from the nephron to the bladder.	Urine flows from the collecting ductules of the nephron into the minor calices. The minor calices empty into the major

calices (wide, cup-shaped structures in the renal sinus). The major calices then empty into the renal pelvis, which is continuous with the ureter.

Identify the labeled structures on the following figure of the kidney and ureter:

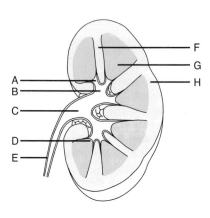

A = Minor calix
B = Major calix
C = Renal pelvis
D = Renal papilla
E = Ureter
F = Renal column
G = Renal pyramid
H = Renal cortex

Where are the adrenal glands located?

The adrenal glands lie on the anteromedial aspect of the superior pole of the kidneys, enveloped by the renal fascia.

Which arteries supply the adrenal gland?

1. The superior adrenal (suprarenal) artery, a branch of the inferior phrenic artery
2. The middle adrenal artery, a branch from the abdominal aorta
3. The inferior adrenal artery, a branch of the renal artery

What is the adrenal gland's venous drainage?

The suprarenal veins drain to the renal vein on the left, but directly to the inferior vena cava on the right. This makes a right adrenalectomy more surgically challenging.

POWER REVIEW

ABDOMEN

Which approximate dermatomal level is represented by the: **Xiphoid process?**	T7
Umbilicus?	T10
Inguinal ligament?	L1
In which direction do the fibers of the external oblique muscle course?	Inferomedially (hands in pockets)
What forms the rectus sheath?	The aponeuroses of the external oblique, internal oblique, and transversus abdominis muscles
What forms the inguinal ligament (Poupart's ligament)?	The folded lower margin of the aponeurosis of the external oblique muscle
The inguinal ligament runs between which 2 structures?	The anterior superior iliac spine and the pubic tubercle
Which major structures pass through the inguinal canal in: **Men?**	The spermatic cord and the ilioinguinal nerve
Women?	The round ligament of the uterus and the ilioinguinal nerve
What are the contents of the spermatic cord?	1. The ductus deferens (vas deferens) 2. The arteries of the ductus deferens (2) 3. The testicular artery 4. The testicular vein (continuous with the pampiniform plexus) 5. The genital branch of the genitofemoral nerve 6. Autonomic nerves 7. Lymphatic vessels draining the testes

Which structures in the abdominal cavity are retroperitoneal?

1. The pancreas
2. Most of the duodenum
3. The ascending and descending colon
4. The aorta
5. The inferior vena cava
6. The kidneys
7. The ureters

Note that the appendix is not retroperitoneal in most patients!

What is the site where the omental bursa (lesser sac) opens into the rest of the peritoneal cavity called?

The epiploic foramen (of Winslow)

What is the name of the right free margin of lesser omentum?

The hepatoduodenal ligament

Which 3 important structures lie in the hepatoduodenal ligament?

1. The common hepatic artery
2. The portal vein
3. The common bile duct

Portal vein **P**osterior, hepatic artery left anterior, bile duct right anterior (LARD: **L**eft **A**rtery, **R**ight **D**uct)

What structure lies in the free margin of the falciform ligament?

The ligamentum teres (the fetal left umbilical vein remnant)

Which structures form the:
  Median umbilical folds?

A remnant of the urachus

  Medial umbilical folds?

Obliterated fetal umbilical arteries

  Lateral umbilical folds?

Inferior epigastric vessels

What is the most posterior recess within the abdominal cavity?

The hepatorenal recess (Morrison's pouch)

What is the extent of the:

    Foregut?

From the oropharynx to the hepatopancreatic ampulla (of Vater)

    Midgut?

From the ampulla of Vater to the distal transverse colon

    Hindgut?

From the distal transverse colon to the anus

What artery supplies the:

    Foregut?

Celiac artery

    Midgut?

Superior mesenteric artery

    Hindgut?

Inferior mesenteric artery

Where does the external iliac artery become the femoral artery?

At the inguinal ligament. Remember: The artery runs *lateral* to the femoral vein.

Which 3 arteries arise from the celiac trunk?

1. Splenic artery
2. Left gastric artery
3. Common hepatic artery

What is the drainage of the:

    Left testicular or ovarian vein?

Into the left renal vein

    Right testicular or ovarian vein?

Directly into the inferior vena cava

Which 2 veins unite to form the portal vein?

The superior mesenteric vein and the splenic vein

Is the left vagus nerve anterior or posterior?

Anterior (Remember **LARP.**)

Which muscle functions as the upper esophageal sphincter?

The cricopharyngeus muscle

Where are the 4 most likely points of constriction or obstruction along the esophagus?

1. Upper esophageal sphincter
2. Aortic arch
3. Left main stem bronchus
4. Lower esophageal sphincter

Which vessels run along the stomach's:

    Lesser curvature?   The right and left gastric arteries

    Greater curvature?   The right and left gastroepiploic arteries

Which arteries supply the duodenum?

1. The superior pancreaticoduodenal arteries (from gastroduodenal artery)
2. The inferior pancreaticoduodenal arteries (from superior mesenteric artery)

Which artery lies directly behind the first portion of the duodenum?

The gastroduodenal artery

Where do the common bile duct and the pancreatic ducts enter the duodenum?

On the posteromedial aspect of the second part of the duodenum (at the hepatopancreatic ampulla of Vater)

What are the major structural differences between the small and large intestine?

The small intestine has plicae circulares, whereas the large intestine has haustra, taenia coli, and appendices epiploicae.

Where do the 3 taeniae coli converge?

At the appendix

Which major vessels run with the pancreas?

The splenic artery (superior) and vein (dorsal)

What organ lies in close contact with the head of the pancreas?

Duodenum

How about the tail?

Spleen

Which structure anatomically divides the liver into right and left lobes?

The falciform ligament

What line divides the liver into functionally independent right and left lobes?

Cantlie's line (imaginary line that runs from gallbladder to inferior vena cava)

Name 4 communications between portal and systemic circulation.

1. Left gastric (coronary) vein with esophageal veins of azygos system
2. Superior rectal vein with middle and inferior rectal veins
3. Paraumbilical veins with epigastric veins of anterior abdominal wall
4. Tributaries of splenic and pancreatic veins with left renal vein

Which structures enter the porta hepatis?

The hepatic artery, portal vein, and bile ducts

Define the cystohepatic (Calot's) triangle.

Bordered by liver superiorly, cystic duct inferiorly, and common hepatic duct medially (contains cystic artery)

Trace the flow of bile from the gallbladder to the intestine.

Gallbladder→cystic duct→common hepatic duct→common bile duct→hepatopancreatic ampulla→second portion of duodenum

The cystic artery normally branches from which artery?

The right hepatic artery

Which 4 abdominal organs are normally in contact with the spleen?

1. The stomach
2. The left kidney
3. The colon
4. The pancreas

Which kidney lies higher?

The left is higher than the right (liver displaces right kidney inferiorly).

What is the adrenal gland's venous drainage?

Renal vein on left, but directly to inferior vena cava on right

10

The Pelvis and Perineum

BONES AND LIGAMENTS

Which 5 bones make up the pelvis?

1. Ilium
2. Ischium
3. Pubis
4. Sacrum
5. Coccyx

Identify the lettered components of the hip bone on the following illustration:

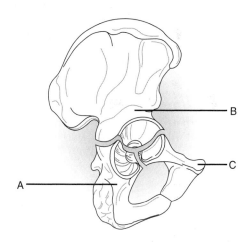

A = Ischium
B = Ilium
C = Pubis

What is the name of the upper ala (wing) of the ilium?

The iliac crest

The most superior point of the iliac crest as palpated posteriorly is located at which vertebral level?	L4
What are the names of the anterior and posterior terminations of the iliac crest?	The anterior superior iliac spine and the posterior superior iliac spine, respectively
What 2 muscles originate from the anterior superior iliac spine?	Sartorius and tensor fascia latae
Which anterior thigh muscle originates from the inner surface of the ilium (iliac fossa)?	The iliacus muscle
Name the sharp bony landmark on the medial side of the ischium.	Ischial spine
What structure attaches to the ischial spine?	The sacrospinous ligament
What are the articulations of the pubis?	The superior ramus articulates with the ilium at the iliopubic eminence; the inferior ramus articulates with the ischial ramus to form the pubic arch.
What is the acetabulum?	A deep hemispherical cup on the lateral surface of the pelvis where the head of the femur articulates with the pelvis. As the acetabulum provides a means for the transfer of weight-bearing forces from the femur to the pelvic girdle, the acetabulum may fracture with high-energy blunt force (e.g., highway-speed motor vehicle accident).
Which bones of the pelvis contribute to the acetabulum?	The ilium, ischium, and pubis

What bony foramen lies between the body of the ischium and the rami of the pubis bones?

The obturator foramen

What structures pass through the obturator foramen?

The obturator nerve, artery, and vein

 What is an obturator hernia?

Herniation of abdominal structures through the obturator foramen; rare condition; usually associated with bowel obstruction due to herniated small bowel or colon

Identify the labeled structures on the following illustration of the hip bone:

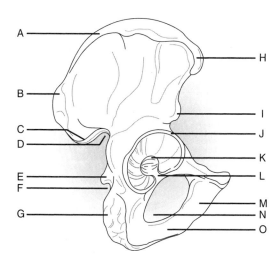

A = Iliac crest
B = Posterior superior iliac spine
C = Posterior inferior iliac spine
D = Greater sciatic notch
E = Ischial spine
F = Lesser sciatic notch
G = Ischial tuberosity

H = Anterior superior iliac spine
I = Anterior inferior iliac spine
J = Acetabular labrum
K = Acetabular fossa
L = Acetabular notch
M = Inferior pubic ramus
N = Obturator foramen
O = Ischial ramus

What are the 4 major articulations of the pelvis?

1. The lumbosacral joint
2. The sacroiliac joints
3. The sacrococcygeal joints
4. The symphysis pubis

Which of these are cartilaginous joints?

The symphysis pubis, the sacrococcygeal joint, and the lumbosacral joint

Which 3 ligaments support the sacroiliac joint?

1. The anterior sacroiliac ligament
2. The posterior sacroiliac ligament
3. The interosseus ligaments

Which 2 ligaments support the symphysis pubis?

The superior pubic ligament and the superior arcuate ligament

Which 3 ligaments support the sacrococcygeal joint?

The anterior, posterior, and lateral sacrococcygeal ligaments

What structure joins the pelvic ring anteriorly?

Pubic symphysis

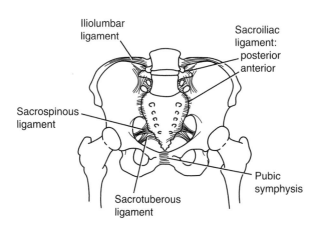

Which ligaments join the sacrum with the ischium?

Sacrotuberous ligament and sacrospinous ligament

Which ligament creates the greater and lesser sciatic foramina?

The sacrotuberous ligament, which extends from the sacrum to the ischial tuberosity

What structures pass out of the pelvic cavity through the greater sciatic foramen?

1. The piriformis muscle
2. The sciatic nerve
3. The internal pudendal artery and vein
4. The pudendal nerve
5. The superior and inferior gluteal vessels and nerves
6. The posterior femoral cutaneous nerve
7. Nerves to the quadratus femoris and obturator internus muscles

What structures pass into the pelvic cavity through the lesser sciatic foramen?

1. The internal pudendal artery and vein
2. The pudendal nerve
3. The nerve to the obturator internus muscle
4. The tendon of the obturator internus muscle

Identify the labeled structures on the following diagram of the bony pelvis:

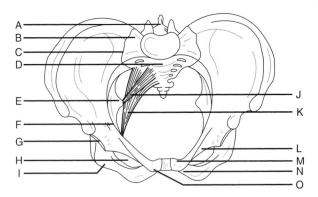

A = First sacral spinous
    process
B = Lateral mass of the
    sacrum
C = Sacroiliac joint
D = Sacral promontory
E = Ischial spine
F = Iliopectineal (arcuate) line
G = Acetabulum

H = Obturator foramen
I = Ischial ramus
J = Sacrospinous ligament
K = Sacrotuberous ligament
L = Superior ramus of the pubis
M = Pubic crest
N = Body of the pubis
O = Pubic tubercle

**Identify the labeled
structures on the following
medial view of the hip bone
and associated structures:**

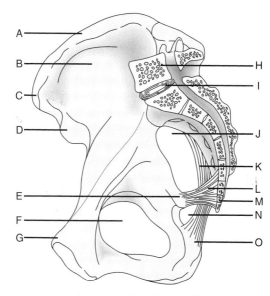

A = Iliac crest
B = Iliac fossa
C = Anterior superior iliac
    spine
D = Anterior inferior iliac spine
E = Ischial spine
F = Obturator foramen
G = Pubic tubercle

H = Vertebral body of L5
I = Lumbosacral joint
J = Greater sciatic foramen
K = Sacrotuberous ligament
L = Sacrospinous ligament
M = Coccyx
N = Lesser sciatic foramen
O = Ischial tuberosity

What is the pelvic inlet?　　The superior rim of the pelvic cavity

**What are the boundaries of
the pelvic inlet:**
　　Posteriorly?　　The sacral promontory

　　Laterally?　　The iliopectineal (arcuate) line

　　Anteriorly?　　The superior margin of the symphysis
　　　　pubis

What is the pelvic outlet?　　The inferior rim of the pelvic cavity

**What are the boundaries of
the pelvic outlet:**
　　Posteriorly?　　The coccyx

　　Laterally?　　The ischial tuberosities

　　Anteriorly?　　The inferior margin of the symphysis pubis

**What is the pelvis major
(false pelvis)?**　　The wide portion of the bony pelvis
　　superior to the pelvic brim (technically,
　　the pelvis major is a part of the abdominal
　　cavity)

**What is the pelvis minor
(true pelvis)?**　　The portion of the bony pelvis inferior to
　　the pelvic brim and superior to the pelvic
　　outlet (Technically, the pelvis minor
　　constitutes the pelvic cavity.)

**List 5 ways the pelvis differs
between women and men.**
1. In women, the bones of the pelvis are
   smaller, lighter, and thinner.
2. In women, the sacrum is broader and
   shorter.

3. In women, the suprapubic arch and the greater sciatic notch are wider.
4. In women, the pelvic inlet is ovoid, while in men, it is heart-shaped.
5. In women, the ischial tuberosities are everted, thereby enlarging the pelvic outlet.

The architecture of the pelvis can vary among individuals. Identify each of the 4 pelvis types on the figure below:

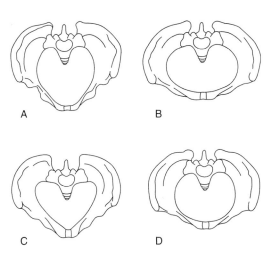

A

B

C

D

A = Anthropoid
B = Platypelloid
C = Android
D = Gynecoid

What is the clinical relevance of these classifications of the bony pelvis?

They help predict whether a vaginal delivery is possible. A gynecoid pelvis is the most favorable. Women with an android pelvis (~25%) may have difficulty with child birth.

What changes in the female pelvis occur during pregnancy?

Pelvic joints and ligaments relax and the coccyx moves posteriorly. All these changes increase the pelvic diameter to facilitate childbirth.

PELVIC MUSCLES AND FASCIAE

What structures form the:
  Anterior pelvic wall?

The pubic bones, the symphysis pubis, the obturator internus muscle, and the obturator internus (parietal pelvic) fascia

  Lateral pelvic wall?

The obturator internus muscle and its associated fascia, the bony pelvis (below the pelvic brim), and the sacrotuberous and sacrospinous ligaments

  Posterior pelvic wall?

The sacrum, coccyx, ligaments, and piriformis muscle

What structures form the floor of the pelvis?

The pelvic and urogenital diaphragms

Weakness ("relaxation") of the pelvic floor musculature is associated with what clinical conditions?

Cystocele (prolapse of bladder into vagina), rectocele (bulging of rectum into vaginal wall), uterine/vaginal prolapse; common in older multiparous (multiple pregnancies) women

Which muscles comprise the pelvic diaphragm?

The levator ani and coccygeus muscles

What is the urogenital diaphragm?

A triangle-shaped sheet of fascia and muscle that spans the space between the ischial spines and pubic symphysis

Which muscles comprise the urogenital diaphragm?

1. The deep transverse perineal muscles, running transversely anteriorly and posteriorly
2. The sphincter urethrae muscle (and the sphincter vaginae muscle, in women)

Identify the labeled structures on the following figure:

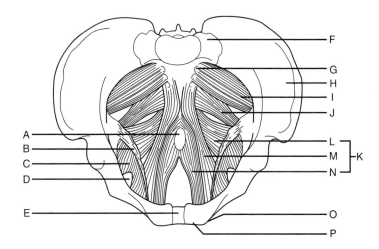

A = Anal canal
B = Tendinous arch of the obturator internus fascia
C = Obturator internus muscle
D = Obturator canal
E = Symphysis pubis
F = Sacrum
G = Sacral foramen
H = Iliac fossa
I = Piriformis muscle
J = Coccygeus muscle
K = Levator ani muscle group
L = Iliococcygeus muscle
M = Pubococcygeus muscle
N = Puborectalis muscle
O = Pubic tubercle
P = Pubic crest

The urogenital diaphragm is pierced by what structure(s):

In males? Membranous urethra

In females? Membranous urethra and vagina

The superior fascia of the urogenital diaphragm is continuous with which fascial layer?	The obturator internus (pelvic parietal) fascia

MUSCLES

Obturator Internus Muscle

Origin?	The pelvic surface of the obturator internus fascia and the surrounding parts of the ilium and pubis
Insertion?	The greater trochanter of the femur (medial part)
Innervation?	The obturator nerve (from the sacral plexus)
Action?	Lateral rotation of the thigh when the hip joint is extended
What structure covers the obturator internus muscle?	The obturator internus fascia
What canal is formed by the obturator internus (pelvic parietal) fascia?	The pudendal canal (transmits the pudendal nerve, artery, and vein)

Piriformis Muscle

Origin?	The anterolateral sacrum and the sacrotuberous ligament
Insertion?	The greater trochanter of the femur (superior border)
Innervation?	Piriformis nerve (branch of the sacral plexus)
Action?	Laterally rotates the thigh when the hip joint is extended and abducts the thigh when the hip joint is flexed

Levator Ani Muscle Group

Which 3 muscles comprise the levator ani muscle group?

1. Puborectalis muscle
2. Pubococcygeus muscle
3. Iliococcygeus muscle

As a group, where do the levator ani muscles:

    Originate?

From the body of the pubis, the tendinous arch of the obturator internus fascia, and the ischial spine

    Insert?

On the perineal body (a fibromuscular mass anterior to the anus), the anococcygeal ligament (the median fibrous intersection of the pubococcygeus muscles), and the walls of the prostate (or vagina), rectum, and anal canal

What is the innervation of the levator ani muscle group?

Spinal nerves S3–S5 and the pudendal nerve

What is the anorectal angle?

The angle between the rectum and the anal canal. Contraction of the puborectalis muscle holds the anorectal junction anteriorly, preventing the passage of feces from the rectum into the anal canal.

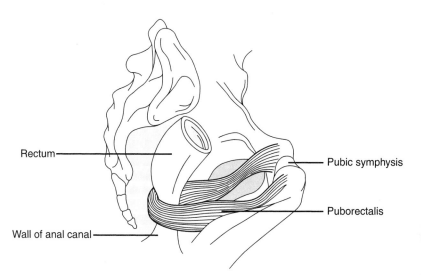

Rectum

Wall of anal canal

Pubic symphysis

Puborectalis

What are the actions of the levator ani muscle group?

1. Supports the pelvic viscera and counteracts increases in the intraabdominal pressure
2. Elevates the floor of the pelvic cavity, assisting in compression of the contents of the pelvic and abdominal cavities

What is the action of the puborectalis muscle in particular?

The puborectalis muscle maintains the anorectal angle, which permits voluntary control of defecation.

Coccygeus Muscle

Origin?

The lateral pelvic surface of the ischial spine and the sacrospinous ligament

Insertion?

The medial and lateral margin of the coccyx and vertebra S5

Innervation?

Branches of spinal nerves S4 and S5

Action?

Assists the levator ani muscle group in supporting the pelvic viscera; supports and pulls the coccyx anteriorly

FASCIAE

The pelvic fascia is consistent with which fascial layer of the abdominal wall?

The transversalis fascia

What are the 2 layers of pelvic fascia?

1. Parietal pelvic fascia (covers the surfaces of the obturator internus, piriformis, levator ani, and coccygeus muscles, both superiorly and inferiorly)
2. Visceral pelvic fascia (binds pelvic organs to each other and to the parietal fascia)

What is the name of the thickening of the obturator internus fascia from which the levator ani muscle group arises?

The tendinous arch

Identify the labeled structures on the following frontal (coronal) section:

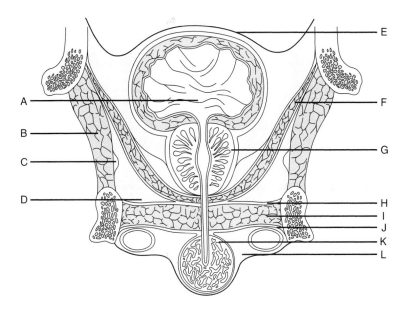

A = Bladder
B = Obturator internus muscle and fascia
C = Pudendal canal
D = Ischioanal (ischiorectal) fossa
E = Peritoneum
F = Pelvic diaphragm
G = Prostate gland
H = Superior fascia of the urogenital diaphragm
I = Urogenital diaphragm
J = Inferior fascia of the urogenital diaphragm (perineal membrane)
K = Bulb of the penis
L = Superficial perineal space

PELVIC VASCULATURE

ARTERIES

The common iliac artery divides into the external and the internal iliac arteries at which vertebral level?	L5–S1
What is the first branch of the internal iliac artery?	The iliolumbar artery
How does the internal iliac artery terminate?	By dividing into anterior and posterior branches
What are the 3 branches from the posterior internal iliac artery?	1. Iliolumbar artery 2. Lateral sacral artery 3. Superior gluteal artery
How does the superior gluteal artery leave the pelvis?	Through the superior part of the greater sciatic foramen (i.e., the part above the piriformis muscle)
What are the 8 branches from the anterior internal iliac artery?	1. Umbilical artery 2. Superior vesical arteries 3. Uterine artery 4. Vaginal artery (in women) or inferior vesical artery (in men) 5. Middle rectal artery 6. Obturator artery 7. Internal pudendal artery 8. Inferior gluteal artery
How does the obturator artery leave the pelvis?	Through the obturator foramen
How does the inferior gluteal artery leave the pelvis?	Through the inferior part of the greater sciatic foramen
What are the 5 branches of the internal pudendal artery?	1. Inferior rectal artery 2. Perineal artery 3. Superficial perineal artery 4. Deep artery of the penis or clitoris 5. Dorsal artery of the penis or clitoris

Identify the labeled arteries and associated structures of the pelvic region on the following sagittal section:

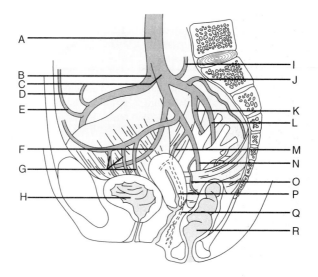

A = Common iliac artery
B = External iliac artery
C = Internal iliac artery
D = Deep circumflex iliac artery
E = Inferior epigastric artery
F = Obturator artery
G = Superior vesical arteries
H = Bladder
I = Iliolumbar artery
J = Lateral sacral artery
K = Superior gluteal artery
L = Inferior gluteal artery
M = Uterine artery
N = Internal pudendal artery
O = Middle rectal artery
P = Uterus
Q = Vaginal artery
R = Rectum

In what clinical setting might the internal iliac artery(s) be intentionally occluded ("embolized") by an interventional radiologist?

Patients with pelvic fractures may present with life-threatening bleeding that is difficult to manage operatively. It is often best treated by internal iliac artery embolization.

What are the potential side effects of internal iliac artery embolization?

Erectile dysfunction and compromise of blood supply to the colon; both more common with bilateral embolization

Where do the testicular/ovarian arteries arise?

Anteriorly from the abdominal aorta, just inferior to the renal arteries

The left common iliac vein is crossed by what major artery?

The *right* common iliac artery. Traumatic injury and bleeding from the left common iliac vein can sometimes only be addressed by intentionally clamping and dividing the right iliac artery to allow access. The artery is reconnected after the vein is repaired.

"Aortic bifurcation"

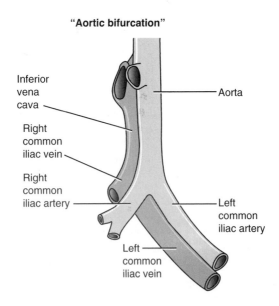

Inferior vena cava

Aorta

Right common iliac vein

Right common iliac artery

Left common iliac artery

Left common iliac vein

What small, unpaired branch courses inferiorly along the sacrum after arising from the aorta near the aortic bifurcation?	The middle (median) sacral artery
What tubular structure crosses directly anterior to the bifurcation of the iliac vessels?	The ureter
What is the relationship between the ureter and the gonadal (ovarian/testicular) vessels?	The gonadal vessels originate medial to the ureter. They cross the ureter *anteriorly* shortly after their origin, after which they run lateral to it.
Why does surgical dissection anterior to the aortic bifurcation (e.g., aortic aneurysm surgery) risk causing erectile dysfunction/impotence?	The superior hypogastric plexus, which contains sympathetic nerves to the pelvic viscera, lies directly anterior to the aortic bifurcation.

LYMPHATICS

What is the drainage of the pelvic lymph nodes (i.e., the external iliac and internal iliac nodes)?	The common iliac nodes
What is the drainage of the perineal lymph nodes?	The perineal lymph nodes drain to the external pudendal nodes, which in turn drain to the superficial inguinal nodes.
What is a pelvic lymph node dissection? For what conditions is it most frequently employed?	Removal of internal, external, and common iliac lymph nodes, all of which travel with their respective arteries; most often indicated for cancer of the prostate or bladder

PELVIC INNERVATION

SOMATIC INNERVATION

Which 2 major nerves are derived from the sacral plexus?	The sciatic nerve (L4–S3) and the pudendal nerve (S2–S4)
The pudendal nerve follows the course of which artery?	The internal pudendal artery
What are the 2 distal branches of the pudendal nerve?	1. The perineal branches 2. The dorsal nerve of the penis (in men) or clitoris (in women)

The pudendal nerve carries which:

Sensory fibers?	Those that supply the external genitalia and perianal region
Motor fibers?	Those that supply the external anal sphincter, muscles associated with ejaculation, and the muscles of the urogenital diaphragm
Which branch of the pudendal nerve supplies the external anal sphincter?	The inferior rectal nerve
Temporary injury to the pudendal nerve may occur with cycling or difficult childbirth. This injury may lead to what symptoms?	Anal/urinary incontinence, sexual dysfunction, loss of sensation in the perianal/genital regions
The sciatic nerve gives off what branches in the pelvis?	None; eventually terminates by branching into the tibial nerve and common peroneal nerve in the lower extremity
What smaller branches of the sacral plexus innervate the gluteal muscles?	Superior and inferior gluteal nerves

What major branch of the *lumbar* plexus arises within the pelvis from spinal nerves L2–L4?

The femoral nerve

Label the following diagram showing the branches of the lumbar plexus:

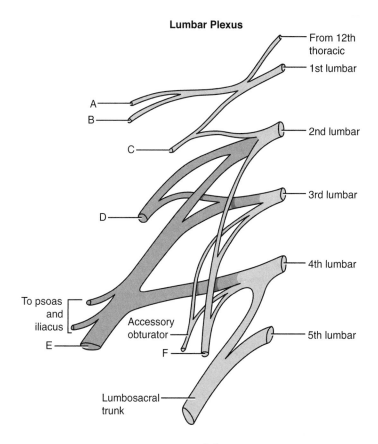

Lumbar Plexus

From 12th thoracic

1st lumbar

2nd lumbar

3rd lumbar

4th lumbar

5th lumbar

To psoas and iliacus

Accessory obturator

Lumbosacral trunk

A = Iliohypogastric nerve
B = Ilioinguinal nerve
C = Genitofemoral nerve
D = Lateral femoral cutaneous nerve
E = Femoral nerve
F = Obturator nerve

How does the femoral nerve exit the pelvis?	Passes inferior to inguinal ligament lateral to femoral artery

AUTONOMIC INNERVATION

Which nerves contribute to the inferior hypogastric plexus?	1. Parasympathetics, via the pelvic splanchnic nerves (derived from spinal segments S2–S4) 2. Sympathetics from the superior hypogastric plexus and pelvic sympathetic trunk
Afferent fibers sensing bladder fullness are carried by which nerves?	The pelvic splanchnic nerves
Visceral efferents to the detrusor muscle and internal sphincter of the bladder are carried by what type of nerves?	Parasympathetics
What type of nerves supply motor innervation to the smooth muscle of the prostate, the seminal vesicle, the ejaculatory ducts, and the ductus deferens (vas deferens)?	Sympathetics
Which autonomic nerves coordinate:	
Erection?	Parasympathetics
Ejaculation?	Sympathetics

PELVIC VISCERA

URETER

What is the orientation of the ureter's origin at the renal hilum, relative to the renal blood vessels?	The ureter lies *posterior* to the renal vessels at its origin (the renal pelvis); from anterior to posterior: Vein, Artery, Pelvis ("VAP")

Describe the course of the ureter from the kidney to the bladder.

The ureter arises from the renal pelvis and descends within the retropritoenum along the anterior surface of the psoas muscle. It crosses the pelvic brim anterior to the bifurcation of the common iliac artery and enters the urinary bladder.

What are the 3 normal anatomic constrictions along the course of the ureter?

1. Renal pelvis (ureteropelvic junction [UPJ])
2. Pelvic brim
3. Entry into bladder (ureterovesical junction [UVJ]).

These 3 sites represent common areas for obstruction, which is most commonly the result of urolithiasis (urinary tract stones).

The ureter is most commonly injured during what surgical procedure?

Hysterectomy (removal of uterus); injury most often occurs in distal ureter, in region of infundibulopelvic ligament; may be due to blind clamping and ties placed to control bleeding

MALE PELVIC VISCERA

Which ducts merge to form the ejaculatory ducts?

The duct of the seminal vesicle and the ductus deferens

Where does the ductus deferens enter the pelvic cavity?

At the deep inguinal ring, lateral to the inferior epigastric artery

How does the ductus deferens travel relative to the ureter?

The ductus deferens travels anterior to the ureter (as the ureter enters the bladder) to join with the duct of the seminal vesicle.

Where do the following structures open into the urethra:
    Ejaculatory ducts?

At the seminal colliculus, lateral and inferior to the prostatic utricle (a remnant of the müllerian duct)

Bulbourethral glands (of Cowper)?	At the spongy urethra
Prostatic ducts?	At the prostatic sinuses (grooves alongside the urethral crest)

Identify the labeled structures on the following schematic figure of the male pelvic viscera:

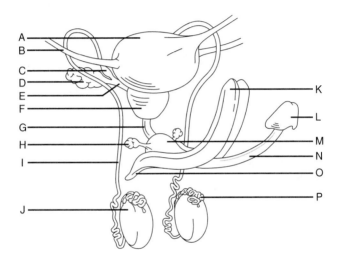

A = Bladder
B = Ureter
C = Ampulla of the ductus deferens
D = Seminal vesicle
E = Ejaculatory duct
F = Prostate gland
G = Urethra (membranous portion)
H = Bulbourethral gland (of Cowper)
I = Ductus deferens (vas deferens)
J = Testis
K = Corpus cavernosum
L = Glans penis
M = Bulb of the penis
N = Corpus spongiosum
O = Crus of the penis
P = Epididymis

 **What is a vasectomy?**

Surgical interruption (tying off or clipping) of the ductus deferens (clinically termed the "*vas* deferens") on both sides using 1 or 2 small scrotal incisions; prevents sperm from entering ejaculate, thereby rendering a male sterile

What are the 3 parts of the male urethra?

Prostatic, membranous, and spongy (penile)

 **What is the most common site of urethral obstruction in males?**

Prostatic urethra; due to benign prostatic hypertrophy (BPH), a very common condition in older males that leads to difficulty with urination

 **What are some physical signs of traumatic urethral injury in a male patient (most often seen with pelvic fracture)?**

Blood at the penile meatus, scrotal ecchymosis (bruising), and a "high-riding" ballotable prostate on digital rectal examination

FEMALE PELVIC VISCERA

What is the name of the recess between the cervix and the wall of the vagina?

The fornix

What are the 4 parts of the uterus?

1. Cervix (external os)
2. Isthmus (internal os)
3. Body
4. Fundus

Name the 3 layers that comprise the wall of the body of the uterus.

1. Perimetrium
2. Myometrium
3. Endometrium (innermost)

 **What is endometriosis?**

The abnormal deposition of tissue resembling uterine endometrium outside the uterus; may occur throughout the pelvis or abdomen; common cause of pelvic pain

Identify the labeled structures on the following figure of the female pelvic viscera:

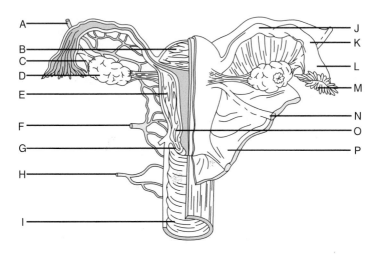

A = Ovarian artery
B = Fundus of the uterus
C = Ovarian ligament
D = Ovary
E = Body of the uterus
F = Uterine artery
G = Cervix
H = Vaginal artery
I = Vagina
J = Fallopian (uterine) tube
K = Ampulla of the fallopian tube
L = Infundibulum of the fallopian tube
M = Fimbriae
N = Round ligament of the uterus
O = Isthmus of the uterus
P = Broad ligament

What is the normal position of the uterus?

Anteflexed (uterine body tips anteriorly relative to the position of the cervix) and anteverted (angled anteriorly relative to the long axis of the vagina)

**What is the normal relation
of the uterus to the bladder?**

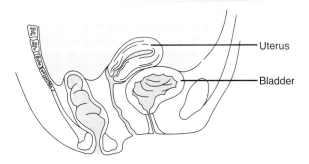

Uterus

Bladder

The uterus normally overlies the bladder
with the fundus positioned anteriorly, at
roughly a right angle to the cervix.

What is the:
    Vesicouterine pouch?

The intraperitoneal space between the
bladder and the uterus

    **Pouch of Douglas
    (rectouterine pouch)?**

The intraperitoneal space
between the uterus and the rectum.
This is a common place for infected
pelvic fluid collections (abscesses) to
form after surgery.

**What are the 5 parts of the
fallopian tube?**

1. Infundibulum
2. Ampulla
3. Isthmus
4. Intramural region
5. Fimbriae

**What is an "ectopic"
pregnancy?**

Implantation of an embryo in a location
other than the uterus. The vast majority
of cases involve implantation in the
fallopian tubes ("tubal pregnancy"). An
ectopic pregnancy may lead to tubal
rupture with associated catastrophic
bleeding.

Identify the 3 mesenteries of the ovary and fallopian tube on the following figure:

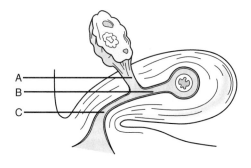

A = Mesovarium
B = Mesosalpinx
C = Mesometrium

What is the name of the double layer of mesentery that runs between the uterus and the pelvic floor/lateral walls?

The broad ligament; functions to maintain the position of the uterus

Which ligament runs anteroinferiorly from the uterus within the broad ligament?

The round ligament of the uterus

Where do the ends of the round ligament of the uterus attach?

1 end attaches to the anterolateral fundic region of the uterus; the other attaches in the labium majus.

 What is "round ligament pain"?

Pelvic pain during pregnancy that is thought to relate to stretching of the round ligament with enlargement of the uterus; no specific treatment

Which 2 ligamentous portions of the broad ligament connect to the ovary?

The suspensory ligament of the ovary and the ovarian ligament proper

How does the suspensory ligament of the ovary differ from the ovarian ligament proper?	The suspensory ligament of the ovary is lateral to the ovary and contains the ovarian vessels, while the ovarian ligament proper is medial to the ovary and connects the ovary and the uterus.
Which ligament attaches the ovary to the wall of the pelvis?	The suspensory ligament of the ovary
Where do the ovarian veins drain to?	The right ovarian vein drains to the inferior vena cava and the left drains to the renal vein (identical to drainage of testicular veins in male).
What ligaments extend from the cervix and lateral parts of the fornix to the pelvic sidewalls?	The transverse cervical ("cardinal") ligaments. These are the chief uterine supports.
What artery runs in the transverse cervical ligament?	The uterine artery
What ligaments connect the posterolateral uterus with the sacrum?	The uterosacral ligaments
How does the female urethra differ from that of the male?	The female urethra has only 1 part (membranous) and is much shorter (~4 cm).
What is the clinical significance of this difference in the female urethra?	Females are *much* more likely to get urinary tract infections, which rely on bacterial ascent through the urethra into the normally sterile bladder.

RECTUM AND ANAL CANAL

How long is the adult rectum?	Approximately 12 cm
Describe the peritoneal coverings of the rectum.	The superior third is covered anteriorly and laterally, the middle third is covered only anteriorly, and the inferior third has

no peritoneal covering owing to peritoneal reflection onto the bladder (in men) or the vagina and uterus (in women).

Which arteries supply the rectum?

1. The superior rectal artery, the direct continuation of the inferior mesenteric artery
2. The middle rectal artery, a branch of the internal iliac artery
3. The inferior rectal artery, a branch of the internal pudendal artery

Identify the arterial supply of the rectum and associated structures on the following figure:

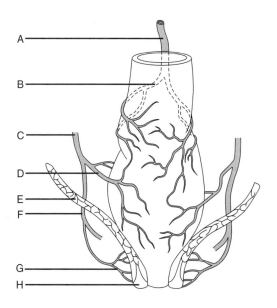

A = Superior rectal artery
B = Right branch of the superior rectal artery
C = Internal iliac artery
D = Middle rectal artery
E = Levator ani muscle group
F = Internal pudendal artery
G = Inferior rectal artery
H = External anal sphincter

Which veins drain the rectum?

The inferior and middle rectal veins (part of the systemic venous system) and the superior rectal vein (part of the portal venous system)

Approximately how long is the anus?

~4 cm

What is the dentate (pectinate) line?

The division between the superior and inferior anal canal

Where is the dentate line located?

Approximately two thirds of the way down the anal canal

What transitions occur at the dentate line?

1. Columnar or cuboidal epithelium above the dentate line changes to stratified squamous epithelium below it.
2. Processes above the dentate line (i.e., hemorrhoids) are painless; those below are painful.
3. Above the dentate line, visceral innervation is supplied via the hypogastric plexus; below the dentate line, somatic innervation is supplied via the pudendal nerve.
4. Venous drainage above the dentate line is via the portal venous system; below the dentate line, venous drainage is to the systemic venous system.
5. Lymphatic drainage above the dentate line is to the internal iliac nodes; below the dentate line, lymphatic drainage is to the superficial inguinal lymph nodes.

What are hemorrhoids?

Dilated vascular channels in the anal hemorrhoidal venous plexus. Risk factors include chronic constipation, low-fiber diet, and pregnancy. They are a common cause of "bright red blood per rectum" and perianal discomfort.

Describe the differences between internal and external hemorrhoids.

Internal hemorrhoids originate from above the dentate line, and are therefore painless; external hemorrhoids have somatic innervation and may be very painful

What are anorectal columns?

Longitudinal folds that extend from the anorectal junction to the dentate line

What are anal sinuses?

Pouchlike recesses between the anorectal columns. The glands of the anal canal open into the anal sinuses.

What is a perianal abscess?

Bacterial infection that originates within the crypts at the base of the anal glands; causes perianal pain/pain with defecation; treated with surgical drainage

PERINEUM

What separates the pelvic cavity from the perineum?

The pelvic floor (aka pelvic diaphragm)

The pelvic floor is composed of what structures?

The levator ani and coccygeus muscles and their associated fascia

What layer of tissue lies just deep to the skin of the perineum?

The superficial perineal fascia

Which layer of abdominal fascia is the superficial perineal fascial (Colles' fascia) continuous with?

The membranous layer of the superficial abdominal fascia (i.e., Scarpa's fascia)

The superficial perineal fascia is continuous with what other tissues?

The tunica dartos in the scrotum and the superficial fascia of the penis

Where is the superficial perineal space?

Between the inferior fascia of the urogenital diaphragm and the superficial perineal fascia

What does it contain?	Superficial transverse perineal, ischiocavernosus, and bulbospongiosus muscles; pudendal nerve; branches of internal pudendal vessels; crus of penis/clitoris; penile/vestibular bulb; perineal body
Where is the deep perineal space?	Between the superior and inferior fascia of the urogenital diaphragm
What does it contain?	Deep transverse perineal and sphincter urethrae muscles; membranous urethra; bulbourethral glands (male); branches of internal pudendal vessels; branches of pudendal nerve
An imaginary line between the ischial tuberosities separates the perineum into which 2 anatomic regions?	1. Anal triangle (posterior) 2. Urogenital triangle (anterior)
What structure lies at the middle of this imaginary line?	The perineal body (aka central tendon); between the external genitalia anteriorly and the anus posteriorly
What is an episiotomy?	Surgical incision of the posterior vaginal wall and perineal body during delivery to prevent jagged tear of perineal muscles during childbirth
Which muscles *insert* into the perineal body?	1. Superficial transverse perineal muscle 2. Deep transverse perineal muscle 3. Levator ani
Which muscles *originate* from the perineal body?	1. Bulbospongiosus muscle 2. External anal sphincter
What are the contents of the urogenital triangle?	The external genitalia and urethra
What are the borders of the anal triangle?	Tip of coccyx and ischial tuberosities

What are the 3 main components of the anal triangle?

1. Anal canal
2. External anal sphincter
3. Ischioanal fossae

Describe the location and shape of the ischioanal fossae (aka ischiorectal fossae).

The ischioanal fossae are wedge-shaped areas located on each side of the anus.

What are the boundaries of the ischioanal fossae?

Base: Skin of perineum
Medial: Levator ani/external anal sphincter, anal canal
Lateral: Obturator internus muscle/fascia
Posterior: Sacrotuberous ligament, gluteus maximus muscle

A perianal abscess can extend into this space, creating an ischiorectal abscess. In severe cases, an abscess may extend to involve the ischiorectal fossa on both sides ("horseshoe abscess").

Which 2 nerves traverse the ischioanal fossa?

1. Inferior rectal nerve
2. Perineal branch of the femoral cutaneous nerve

What else does the ischioanal fossa contain?

Ischiorectal fat and the inferior rectal artery and vein

What canal runs in the lateral aspect of the ischioanal space?

The pudendal canal (Alcock's canal); contains pudendal nerve and vessels and dorsal nerve of penis/clitoris

What are the 3 major parts of the external anal sphincter?

Subcutaneous, superficial, and deep

With which muscle does the deep part of the external anal sphincter blend?

The puborectalis muscle (part of levator ani)

Which nerve innervates the external anal sphincter?

The inferior rectal nerve (branch of pudendal nerve)

How does control of the external anal sphincter differ from that of the internal anal sphincter?

Control of the external anal sphincter is voluntary, whereas control of the internal anal sphincter is involuntary (parasympathetic).

What causes relaxation of the internal anal sphincter?

Distension of the rectum

Which sphincter must relax for defecation to occur?

Both

MALE PERINEUM

Penis

What are the 3 anatomic divisions of the penis, and which structures comprise each?

1. Root of the penis, composed of the 2 crura and the bulb of the penis
2. Body of the penis, composed of the corpus spongiosum and the 2 corpora cavernosa
3. Glans of the penis, composed of the terminal corpus spongiosum

Describe the origin of each corpa cavernosa.

From the ischiopubic ramus, via the "crus" of the penis

Where do the ischiocaver-nosus muscles:
    Originate?

The ischial tuberosities

    Insert?

Penile crura

Where does the bulbospon-giosus muscle:
    Originate?

From the perineal body

    Insert?

The bulb, dorsum, and side of the penis

What is the action of the ischiocavernosus and bulbospongiosus muscles?

They retard venous return, thereby permitting erection to occur, and facilitate the expulsion of ejaculate at the time of ejaculation.

 What is a priapism?

A prolonged, painful erection; common complication of in patients with sickle cell anemia (due to venous outflow obstruction)

What is the innervation of the ischiocavernosus and bulbospongiosus muscles?

The perineal branch of the pudendal nerve

What arteries supply the penis?

The dorsal and deep arteries of the penis, branches of the internal pudendal artery

Between which layers of the penis are the dorsal arteries located?

Between the tunica albuginea (a fascial layer that envelops the corpus cavernosum and corpus spongiosum) and Buck's (deep) fascia

Where are the deep arteries of the penis located?

Within the corpora cavernosa

Where is the deep dorsal vein of the penis located?

Between the tunica albuginea and Buck's (deep) fascia (like the dorsal arteries)

Identify the labeled structures on the following figure of the penis:

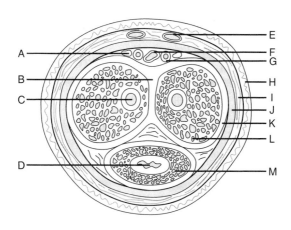

A = Dorsal nerve of the penis
B = Septum penis
C = Deep artery of the penis
D = Urethra
E = Superficial dorsal vein
F = Deep dorsal vein
G = Dorsal artery of the penis

H = Skin
I = Superficial fascia
J = Buck's (deep) fascia
K = Tunica albuginea
L = Corpus cavernosum
M = Corpus spongiosum

Scrotum and Testes

Name the layers around the testicles, from superficial to deep.

Scrotal skin, superficial (dartos) fascia, external spermatic fascia, cremaster muscle/fascia, internal spermatic fascia, tunica vaginalis, tunica albuginea

What is the nature and action of the dartos fascia?

Composed mostly of smooth muscle; functions to regulate the temperature for spermiogenesis by retracting and relaxing the scrotum

What are the 2 layers of the tunica vaginalis?

An inner visceral layer, which covers the testis, and a parietal layer

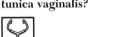 **What is a patent processus vaginalis?**

During development, the testicle descends into the scrotum from the abdomen via the processus vaginalis, which normally closes. When this opening remains open ("patent"), it predisposes the later development of an indirect inguinal hernia.

What lies directly internal to the visceral layer of the tunica vaginalis?

The tunica albuginea

 What is a hydrocele?

Fluid within the tunica vaginalis; presents as painless scrotal swelling; can be congenital (peritoneal fluid enters tunica via a patent processus vaginalis) or can result from tumor, trauma, or inflammation

What is cryptorchidism?

The failure of normal testicular descent into the scrotum; testicle remains in abdomen or in inguinal canal; surgically repositioned if still undescended by age 1

What structure normally anchors the testicle to the scrotum?

The gubernaculum testis. Its abnormal development predisposes testicular torsion, twisting of the testicle around its blood supply (a surgical emergency!).

Identify the lettered structures on the following diagram:

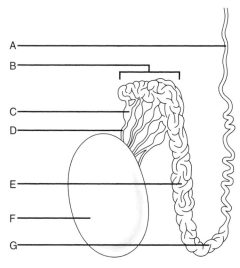

A = Ductus deferens
B = Head of the epididymis
C = Lobules of the epididymis
D = Efferent ductules of the testis
E = Body of the epididymis
F = Testis
G = Tail of the epididymis

What is the innervation of the scrotum?

The scrotum is innervated by the ilioinguinal, genitofemoral, perineal, and posterior femoral cutaneous nerves.

What is the lymphatic drainage of the:
  Scrotum?

Superficial inguinal lymph nodes

  Testis?

The aortic and retroperitoneal lymph nodes

Into which vein does the testicular vein drain on the:

    Left? — The left renal vein

    Right? — Inferior vena cava

What is a varicocele? — Abnormal dilation of the testicular venous plexus within the scrotum; causes painless swelling above testicle; may be due to venous outflow obstruction or be "primary"

FEMALE PERINEUM

What structure is the female homolog of the male scrotum? — The labia majora

Identify the labeled structures on the following figure of the external female genitalia:

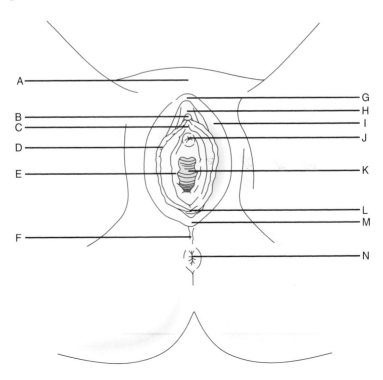

A = Mons pubis
B = Glans clitoris
C = Frenulum of the clitoris
D = Labium minus (labia minora, plural)
E = Hymen (ruptured)
F = Perineal body (central perineal tendon)
G = Anterior labial commissure
H = Prepuce (hood of the clitoris)
I = Labium majus (labia majora, plural)
J = External urethral orifice
K = Vaginal orifice
L = Frenulum of the labia minora
M = Posterior labial commissure
N = Anus

The labia minora form folds to cover what anterior structure?

The clitoris

The urethral orifice is accompanied by what glands?

Paraurethral (Skene's) glands; equivalent to male prostate

What glands lie deep to the posterior vestibule (opening) of the vagina?

Greater vestibular (Bartholin's) glands; secrete a lubricating mucus into the vestibule of the vagina

Describe the structure of the clitoris.

2 crura originate from the ischiopubic rami and become the corpora cavernosa; the corpora unite to form the body of the clitoris, which terminates as the glans.

POWER REVIEW

PELVIS

Which ligament creates the sciatic foramen?	The sacrotuberous ligament
Which ligament divides the sciatic foramen into the greater and lesser sciatic foramina?	The sacrospinous ligament
What structures pass out of the pelvic cavity through the greater sciatic foramen?	1. The piriformis muscle 2. The sciatic nerve 3. The internal pudendal artery and vein 4. The pudendal nerve 5. The superior and inferior gluteal vessels and nerves 6. The posterior femoral cutaneous nerve 7. Nerves to the quadratus femoris and obturator internus muscles
What structures pass into the pelvic cavity through the lesser sciatic foramen?	1. The internal pudendal artery and vein 2. The pudendal nerve 3. The nerve to the obturator internus muscle 4. The tendon of the obturator internus muscle
Which muscles comprise the pelvic diaphragm?	The levator ani and coccygeus muscles
Which muscles comprise the urogenital diaphragm?	1. The deep transverse perineal muscles, running transversely anteriorly and posteriorly 2. The sphincter urethrae muscle (and the sphincter vaginae muscle, in women)
The urogenital diaphragm is pierced by what structure(s): **In males?**	Membranous urethra
In females?	Membranous urethra and vagina

Which 3 muscles comprise the levator ani group?	1. Iliococcygeus muscle 2. Pubococcygeus muscle 3. Puborectalis muscle
From which artery does the testicular or ovarian artery arise on the:	
Left?	The aorta
Right?	The aorta
Into which vein does the testicular or ovarian vein drain on the:	
Left?	The left renal vein
Right?	Directly into the inferior vena cava
Which arteries supply the rectum?	1. The superior rectal artery, the direct continuation of the inferior mesenteric artery 2. The middle rectal artery, a branch of the internal iliac artery 3. The inferior rectal artery, a branch of the internal pudendal artery
Which veins drain the rectum?	The inferior, middle, and superior rectal veins
Which rectal veins are:	
Systemic?	The inferior and middle rectal veins
Portal?	The superior rectal vein
Describe the peritoneal coverings of the rectum.	The superior third is covered anteriorly and laterally, the middle third is covered only anteriorly, and the inferior third has no peritoneal covering.
The pudendal nerve carries which:	
Sensory fibers?	Those that supply the external genitalia and perianal region

Motor fibers?

Those that supply the external anal sphincter, muscles associated with ejaculation, and the muscles of the urogenital diaphragm

What is the origin and path of the pudendal nerve in the pelvis?

The pudendal nerve, a branch of the sacral plexus, accesses the gluteal region by exiting the pelvis through the greater sciatic foramen; it then reenters the pelvis through the lesser sciatic foramen to travel in the pudendal canal.

Which 3 nerves arise from the pudendal nerve?

1. Inferior rectal nerve
2. Perineal nerve
3. Dorsal nerve of the penis (in men) or clitoris (in women)

What is the orientation of the ureter's origin at the renal hilum, relative to the renal blood vessels?

The ureter lies *posterior* to the renal vessels at its origin (the renal pelvis); from anterior to posterior: Vein, Artery, Pelvis ("VAP")

What are the 3 normal anatomic constrictions along the course of the ureter?

1. Renal pelvis (ureteropelvic junction [UPJ])
2. Pelvic brim
3. Entry into bladder (ureterovesical junction [UVJ])

What are the 3 parts of the male urethra?

Prostatic, membranous, and spongy (penile)

What are the 4 parts of the uterus?

1. Cervix (external os)
2. Isthmus (internal os)
3. Body
4. Fundus

What are the 5 parts of the fallopian tube?

1. Infundibulum
2. Ampulla
3. Isthmus
4. Intramural region
5. Fimbriae

What is the name of the double layer of mesentery that runs between the uterus and the pelvic floor/lateral walls?

The broad ligament

An imaginary line between the ischial tuberosities separates the perineum into which 2 anatomic regions?

1. Anal triangle (posterior)
2. Urogenital triangle (anterior)

What structure lies at the middle of this imaginary line?

The perineal body (aka central tendon)

Which muscles *insert* into the perineal body?

1. Superficial transverse perineal muscle
2. Deep transverse perineal muscle
3. Levator ani

Which muscles *originate* from the perineal body?

1. Bulbospongiosus muscle
2. External anal sphincter

What are the 3 major parts of the external anal sphincter?

Subcutaneous, superficial, and deep

With which muscle does the deep part of the external anal sphincter blend?

The puborectalis muscle (part of levator ani)

Which nerve innervates the external anal sphincter?

The inferior rectal nerve (branch of pudendal nerve)

How does control of the external anal sphincter differ from that of the internal anal sphincter?

Control of the external anal sphincter is voluntary, whereas control of the internal anal sphincter is involuntary (parasympathetic).

Name the layers around the testicles, from superficial to deep.

Scrotal skin, superficial (dartos) fascia, external spermatic fascia, cremaster muscle/fascia, internal spermatic fascia, tunica vaginalis, tunica albuginea

What structure normally anchors the testicle to the scrotum?

The gubernaculum testis

11   The Lower Extremity

BONES

What are the 4 regions of the lower limb, and which bones are found in each region?

1. **Hip:** Ilium, ischium, and pubis
2. **Thigh:** Femur and patella
3. **Leg:** Tibia and fibula
4. **Foot:** Tarsal bones, metatarsal bones, and phalanges

THIGH

Identify the labeled structures on the following figure of the posterior right hip and thigh:

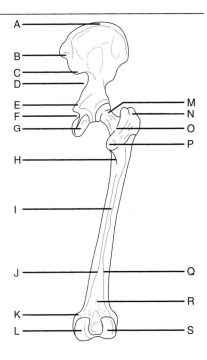

A = Iliac crest
B = Posterior superior iliac spine
C = Posterior inferior iliac spine
D = Greater sciatic notch
E = Ischial spine
F = Lesser sciatic notch
G = Ischial tuberosity
H = Pectineal line
I = Linea aspera

J = Medial supracondylar line
K = Adductor tubercle
L = Medial femoral condyle
M = Neck of the femur
N = Greater trochanter of the femur
O = Intertrochanteric crest
P = Lesser trochanter of the femur
Q = Lateral supracondylar line
R = Popliteal surface
S = Lateral femoral condyle

Femur

Describe the proximal end of the femur.

It consists of a head, neck, and the greater and lesser trochanters.

What structure separates the greater trochanter of the femur from the skin?

The trochanteric bursa

What structure attaches to the intertrochanteric line?

The iliofemoral ligament, a substantial ligament that contributes to the capsule of the hip joint

What is the intertrochanteric crest?

A prominent ridge that unites the 2 trochanters posteriorly

What is the quadrate tubercle?

A prominence on the intertrochanteric crest where the quadratus femoris muscle attaches

What is the fovea capitis?

A pit located roughly in the center of the femoral head, where the ligament of the femoral head attaches

What is the linea aspera?

A ridge that runs down the posterior midline of the femoral shaft

What becomes of the lateral and medial lips of the aspera proximally?

The lateral lip becomes the gluteal linea tuberosity (the insertion site of the gluteus maximus muscle) and the medial lip continues as the spiral line to join the iliotibial line.

What connects the lesser trochanter with the linea aspera?

The pectineal line

Which structure attaches to the pectineal line?

The tendon of the pectineus muscle

Describe the distal end of the femur.

The distal end of the femur has medial and lateral condyles, which articulate with the tibia and patella to form the knee joint.

Describe the articular surfaces of the distal femur:
    Anteriorly

The medial and lateral femoral condyles articulate with the patella.

    Posteriorly

The medial and lateral condyles of the femur articulate with the medial and lateral tibial condyles of the tibia.

Patella

What type of bone is the patella?

A sesamoid bone (the largest in the body)

Where is the patella located?

Within the tendon of the quadriceps femoris muscle in the anterior knee

Name 2 functions of the patella.

1. Attaches the quadriceps femoris tendon to the tibial tuberosity via the patellar ligament
2. Increases the power of the quadriceps femoris muscle by increasing its leverage

What are the attachments of the patellar ligament?

Inferior patella and tibial tuberosity (often regarded as a distal continuation of the rectus femoris tendon)

What is patellar tendonitis?

Inflammation and degeneration at the insertion of the patellar ligament (not a tendon!); leads to pain at tibial tuberosity; usually related to overuse/athletic activity

Identify each part of a long bone, as exemplified by the femur below:

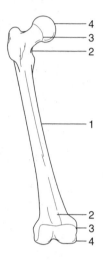

1. Diaphysis (shaft)
2. Metaphysis (neck)
3. Physis
4. Epiphysis (end)

LEG

Identify the labeled landmarks on the following anterior view of the leg:

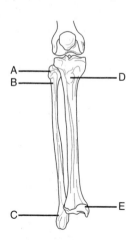

A = Head of the fibula
B = Neck of the fibula
C = Lateral malleolus (of the fibula)
D = Tibial tuberosity
E = Medial malleolus (of the tibia)

What is the name of the well-formed ridge on the posterior tibial surface?

The soleal line

Which bones make up the medial and lateral malleolus of the ankle joint?

The inferomedial process of the tibia forms the medial malleolus. The inferior process of the fibula forms the lateral malleolus

What is the primary weight bearing bone in the leg?

The tibia

With which bone of the foot does the tibia articulate?

The talus (1 of the 7 tarsal bones)

FOOT

Identify the labeled structures on the following posterior view of the leg and the plantar aspect of the foot:

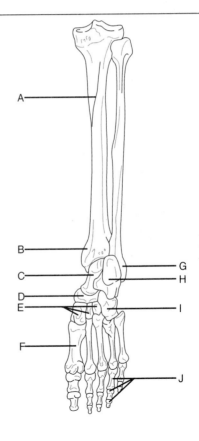

A = Soleal line (of the tibia)
B = Medial malleolus (of the tibia)
C = Talus
D = Navicular bone
E = Cuneiform bones
F = Metatarsal bone (1 of 5)
G = Lateral malleolus (of the fibula)
H = Calcaneus
I = Cuboid bone
J = Phalanges (3 of 14)

Identify the labeled structures on the following lateral view of the foot:

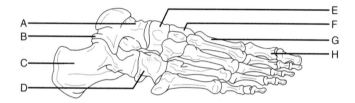

A = Talus
B = Posterior process of talus
C = Calcaneus
D = Cuboid bone
E = Navicular bone
F = Cuneiform bone (1 of 3)
G = Metatarsal bone (1 of 5)
H = Phalanx (1 of 14)

Name the 7 tarsal bones.

1. Talus
2. Calcaneus
3. Cuboid bone
4. Navicular bone
5. Cuneiform bones (3)

What is the largest and most posterior tarsal bone?

The calcaneus

What structure inserts into the posterior surface of the calcaneus?

The tendo calcaneus (Achilles tendon)

What is the sustentaculum tali?

A shelflike medial projection of the calcaneus that supports the talus

What structure lies in the large groove under the sustentaculum tali?

The flexor hallucis longus tendon

The talus articulates with which 2 tarsal bones?

The navicular bone and the calcaneus

The calcaneus articulates with which 2 tarsal bones?

The talus and the cuboid bone

The navicular bone articulates with which 5 tarsal bones?

The talus, the cuboid bone, and the 3 cuneiform bones

Name the 4 intertarsal joints.

1. Talocalcaneonavicular
2. Talocalcaneal
3. Calcaneocuboid
4. Cuneonavicular

Which movements occur around the intertarsal joints?

Inversion and eversion of the hindfoot

What structure passes through the large groove on the underlying surface of the cuboid bone?

The peroneus longus tendon

Describe the articulation of the metatarsal bones with the tarsal bones.

The first 3 metatarsal bones articulate with the cuneiform bones. The fourth and fifth metatarsal bones articulate with the cuboid bone (i.e., the cuboid bone lies laterally).

Which metatarsal bone has 2 articular facets for articulation with sesamoid bones?

The first metatarsal bone

Which muscle inserts into a tuberosity at the base of the fifth metatarsal bone?

The peroneus brevis muscle

What is the clinical significance of the insertion of the peroneus brevis muscle into the tuberosity of the fifth metatarsal bone?

An avulsion fracture (chip fracture) of the tuberosity of the fifth metatarsal bone can occur during inversion injuries of the ankle.

GLUTEAL REGION

Name the 9 muscles of the gluteal region.

1. Gluteus maximus muscle
2. Gluteus minimus muscle
3. Gluteus medius muscle
4. Tensor fasciae latae muscle
5. Piriformis muscle
6. Obturator internus muscle
7. Gemellus inferior muscle
8. Gemellus superior muscle
9. Quadratus femoris muscle

Which artery supplies the:

Gluteus maximus muscle?

The inferior gluteal artery

Gluteus medius and minimus muscles?

The superior gluteal artery

Which of the gluteal muscles is considered the largest single muscle in the body?

The gluteus maximus muscle

Name the 3 bursae that separate the gluteus maximus muscle from the underlying structures.

1. Trochanteric bursa
2. Gluteofemoral bursa
3. Ischial bursa

What is the origin of the gluteus maximus muscle?

The ilium, sacrum, coccyx, and sacrotuberous ligament

Insertion?

The gluteal tuberosity (i.e., the lateral lip of the linea aspera of the femur) and the iliotibial tract (the portion of the fascia lata that extends from the iliac crest to the tibia)

Innervation?

The inferior gluteal nerve

Action?	Chief extensor of the thigh and a lateral (external) rotator of the hip joint

GLUTEUS MINIMUS MUSCLE

Origin?	The ilium, between the anterior and posterior gluteal lines
Insertion?	The anterior aspect of the greater trochanter of the femur
Innervation?	The superior gluteal nerve
Action?	Abducts and medially (internally) rotates the thigh

GLUTEUS MEDIUS MUSCLE

Origin?	The ilium, between the anterior and posterior gluteal lines
Insertion?	The greater trochanter of the femur
Innervation?	The superior gluteal nerve
Action?	Abducts the thigh and, along with the gluteus minimus muscle, maintains stability of the pelvis during ambulation

TENSOR FASCIAE LATAE MUSCLE

Origin?	The anterior superior iliac spine and the external lip of the iliac crest
Insertion?	The lateral tibial condyle (via the iliotibial tract)
Innervation?	The superior gluteal nerve
Action?	Abducts, medially (internally) rotates, and flexes the thigh

PIRIFORMIS MUSCLE

Origin?	The pelvic surface of the anterior sacrum and the sacrotuberous ligament

Insertion?	The superior greater trochanter, in the piriformis fossa
Innervation?	The piriformis nerve (ventral rami of S1–S2)
Action?	Rotates the thigh laterally (externally)

OBTURATOR INTERNUS MUSCLE

Origin?	The ischiopubic rami and the obturator membrane
Insertion?	The medial greater trochanter of the femur
Innervation?	The nerve to the obturator internus
Action?	Abducts and laterally (externally) rotates the thigh

GEMELLI MUSCLES

Origin?	**Gemellus superior:** The ischial spine **Gemellus inferior:** The ischial tuberosity
Insertion?	Medial greater trochanter (common insertion with the obturator internus tendon)
Innervation?	**Gemellus superior:** The nerve to the obturator internus **Gemellus inferior:** The nerve to the quadratus femoris
Action?	Abducts and externally rotates the thigh

QUADRATUS FEMORIS MUSCLE

Origin?	The ischial tuberosity
Insertion?	The quadrate tubercle of the intertrochanteric crest
Innervation?	The nerve to the quadratus femoris
Action?	Rotates the thigh laterally (externally)

HIP REGION

What type of joint is the hip joint?	A synovial ball-and-socket joint
Which bones articulate at the hip joint?	The head of the femur and the pelvic bones (at the acetabulum)

Hip dislocation most commonly occurs in which direction?

In the vast majority of cases, the femoral head dislocates in a posterior position relative to the acetabulum. The affected leg appears shortened, internally rotated, and adducted.

Which muscle is the major flexor at the hip joint?	The iliopsoas muscle, 1 of the anterior thigh muscles

Name the external rotators of the hip.

Remember: "**P**lay **G**olf **O**r **G**o **O**n **Q**uaaludes"
Piriformis
Gemellus superior
Obturator internus
Gemellus inferior
Obturator externus
Quadratus femoris

Name the 5 ligaments that are associated with the hip joint.

1. Iliofemoral ligament
2. Ischiofemoral ligament
3. Pubofemoral ligament
4. Transverse acetabular ligament
5. Ligament capitis femoris

Which of these ligaments is the strongest and most important clinically?	The iliofemoral ligament

THIGH

What is the thigh?	The limb segment between the hip and knee joints

MUSCLES

Name the 3 muscle compartments of the thigh.	Posterior, medial, and anterior

Posterior Thigh Muscles

List the 4 muscles of the posterior thigh compartment.

1. Semimembranous muscle
2. Semitendinosus muscle
3. Biceps femoris muscle (long and short heads)
4. Adductor magnus muscle (hamstring part)

What are the "hamstring" muscles?

The semimembranosus muscle, the semitendinosus muscle, the long head of the biceps femoris muscle, and the adductor magnus muscle (hamstring part)

Identify the muscles and related structures on the following figure of the gluteal region and posterior thigh:

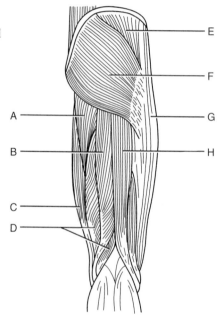

A = Adductor magnus muscle (hamstring part)
B = Semitendinosus muscle
C = Gracilis muscle
D = Semimembranosus muscle
E = Gluteus medius muscle
F = Gluteus maximus muscle
G = Iliotibial tract
H = Biceps femoris muscle

What is the common origin of the posterior thigh muscles?	The ischial tuberosity
What is the common innervation of the posterior thigh muscles?	The tibial division of the sciatic nerve

Semimembranosus Muscle

Origin?	The ischial tuberosity
Insertion?	The posterior part of the medial tibial condyle
Innervation?	The sciatic nerve (tibial division)
Action?	Extends the thigh and flexes the leg

Semitendinosus Muscle

Origin?	The ischial tuberosity
Insertion?	The proximal medial tibia (pes anserinus)
Innervation?	The sciatic nerve (tibial division)
Action?	Extends the thigh and flexes the leg

Biceps Femoris Muscle

Origin?	**Long head:** The ischial tuberosity **Short head:** The lateral lip of the linea aspera and the lateral supracondylar line of the femur
Insertion?	The head of the fibula and lateral proximal tibia
Innervation?	The sciatic nerve (The tibial division innervates the long head and the common peroneal division innervates the short head.)
Action?	Flexes the leg and extends the thigh

Which of the medial thigh muscles contributes to the action of the hamstrings?	The adductor magnus muscle has 2 portions with separate insertions and innervations, 1 of which contributes to the action of the hamstrings (flex the leg).

Adductor Magnus Muscle (hamstring part)

Origin?	The ischial tuberosity
Insertion?	The adductor tubercle of the medial epicondyle of the femur
Innervation?	The sciatic nerve (tibial division)
Action?	Adducts and extends the thigh

Medial Thigh Muscles

List the 6 muscles of the medial thigh compartment.	1. Pectineus muscle 2. Adductor longus muscle 3. Adductor magnus muscle (adductor part) 4. Adductor brevis muscle 5. Gracilis muscle 6. Obturator externus muscle
Where do the 3 "adductors" insert?	On the linea aspera of the femur
Which medial thigh muscles other than the 3 "adductors" contribute to thigh adduction?	The gracilis muscle and the pectineus muscle
Which of these 2 muscles is the most medial?	The gracilis muscle (the pectineus muscle is more lateral)
Of the 5 muscles in the medial thigh group that contribute to adduction, which can be removed without noticeable loss of function?	The gracilis muscle (often transplanted with its nerves and blood supply to replace a damaged muscle)

Pectineus Muscle

Origin?	The pecten pubis (pectineal line of the pubis)

Insertion?	The pectineal line of the femur
Innervation?	The femoral nerve
Action?	Adducts and flexes the thigh

Adductor Longus Muscle

Origin?	The body of the pubis (inferior to the pubic crest)
Insertion?	The middle third of the linea aspera of the femur
Innervation?	The obturator nerve
Action?	Adducts the thigh

Adductor Magnus Muscle (adductor part)

Origin?	The pubic and ischial rami
Insertion?	The gluteal tuberosity, the medial linea aspera, and the supracondylar line of the femur
Innervation?	The obturator nerve
Action?	Adducts and flexes the thigh

Adductor Brevis Muscle

Origin?	The body and inferior ramus of the pubis
Insertion?	The pectineal line and the proximal part of the linea aspera of the femur
Innervation?	The obturator nerve
Action?	Adducts and flexes the thigh

Gracilis Muscle

Origin?	The body and inferior ramus of the pubis
Insertion?	The proximal medial tibia (pes anserinus)

Innervation?   The obturator nerve

Action?   Adducts the thigh and flexes and
                                        medially (internally) rotates the leg

Obturator Externus Muscle

Origin?   The obturator membrane and the
                                        margins of the obturator foramen

Insertion?   The trochanteric fossa of the femur

Innervation?   The obturator nerve

Action?   Laterally (externally) rotates the thigh

Anterior Thigh Muscles

List the 3 muscles of the   1. Iliopsoas muscle
anterior compartment of the   2. Sartorius muscle
thigh.   3. Quadriceps femoris muscle

Iliopsoas Muscle

Which 2 muscles comprise   The iliacus and the psoas major muscles
the iliopsoas muscle?

Iliacus Muscle

Origin?   The iliac crest, iliac fossa, sacral ala, and
                                        the anterior sacroiliac ligament

Insertion?   Inferior to the lesser trochanter of the
                                        femur, via the iliopsoas tendon

Innervation?   The femoral nerve

Action?   Flexes the hip and stabilizes the hip
                                        joint in conjunction with the psoas
                                        major muscle

Psoas Major Muscle

Origin?   The bodies and intervertebral disks of
                                        vertebrae T12–L5

Insertion?   The lesser trochanter of the femur (via
                                        the iliopsoas tendon)

Innervation?	The ventral primary rami of spinal cord levels L1–L3
Action?	Flexes the hip and stabilizes the hip joint in conjunction with the iliacus muscle

Sartorius Muscle

Origin?	The anterior superior iliac spine and from the ilial notch, below the anterior superior iliac spine
Insertion?	The proximal medial tibia (pes anserinus)
Innervation?	The femoral nerve
Action?	Flexes, abducts, and laterally (externally) rotates the thigh at the hip and flexes the knee

Quadriceps Femoris Muscle

Which 4 muscles contribute to the quadriceps femoris muscle?	1. Rectus femoris muscle 2. Vastus lateralis muscle 3. Vastus medialis muscle 4. Vastus intermedius muscle
State where each of the 4 muscles that contribute to the quadriceps femoris muscle originates:	
Rectus femoris muscle?	The anterior inferior iliac spine and the ilial ala (superior to the acetabulum)
The vastus lateralis muscle?	The greater trochanter and lateral lip of the linea aspera of the femur
Vastus medialis muscle?	The intertrochanteric line and the medial lip of the linea aspera of the femur
Vastus intermedius muscle?	The anterior and lateral surfaces of the body of the femur
Where does the quadriceps femoris muscle insert?	The quadriceps tendon inserts into the superior border of the patella and continues as the patellar ligament to insert on the tibial tuberosity.

Which nerve innervates the quadriceps femoris muscle?	The femoral nerve
What is the primary action of the quadriceps femoris muscle?	It extends the leg at the knee joint.
What other action does the rectus femoris have?	It contributes to flexion of the thigh.

FASCIAE

Name the deep fascia of the thigh.	Fascia lata
What is the name of the thick, strong, lateral portion of the fascia lata in the thigh?	The iliotibial tract
Where does the iliotibial tract originate?	On the tubercle of the iliac crest
Where does it insert?	On the lateral condyle (of the tibia)
What is the function of the iliotibial tract?	It acts as tendon for the tensor fasciae lata muscle and contributes to the tendon of the gluteus maximus muscle, thereby steadying the trunk on the thighs.
What structures are housed within the superficial fascia of the thigh (tela subcutanea)?	The cutaneous vessels and the superficial inguinal lymph nodes

VASCULATURE

Arteries

Which arteries represent the primary arterial supply to the thigh?	The superficial and deep femoral arteries
Describe the course of the deep femoral artery.	The deep femoral artery (aka profunda femoris artery) arises posteriorly from the common femoral artery. It passes between the pectineus and adductor longus muscles

and descends posterior to the latter muscle, giving off perforating branches that supply the adductor magnus muscle and the posterior thigh muscles.

Identify the labeled arteries of the lower limb on the following figure:

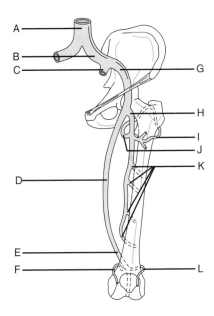

A = Aorta
B = Common iliac artery
C = Internal iliac artery
D = Superficial femoral artery
E = Popliteal artery
F = Superior medial geniculate artery
G = External iliac artery
H = Deep femoral artery (profunda femoris)
I = Lateral circumflex femoral artery
J = Medial circumflex femoral artery
K = Perforating arteries
L = Superior lateral geniculate artery

Which branches of the deep femoral artery supply the head and neck of the femur and the muscles on the lateral side of the thigh?

The medial and lateral circumflex femoral arteries

Where does the superficial femoral artery terminate?

It becomes the popliteal artery at the adductor hiatus (an opening in the adductor magnus muscle).

Define the arterial anatomy of the lower extremity on the figure below:

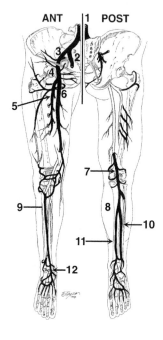

1. Aorta
2. Internal iliac (hypogastric)
3. External iliac
4. Common femoral artery
5. Profundi femoral artery
6. Superficial femoral artery (SFA)
7. Popliteal artery
8. Trifurcation
9. Anterior tibial artery
10. Peroneal artery
11. Posterior tibial artery
12. Dorsalis pedis artery

Veins

What are the 2 most important tributaries of the femoral vein?

The great saphenous vein and the deep femoral vein

What is the longest vein in the body?

The great saphenous vein

Name the 5 major tributaries to the great saphenous vein.

1. Superficial circumflex iliac vein
2. Superficial epigastric vein in the thigh.
3. External pudendal vein
4. Lateral femoral cutaneous vein
5. Anterior femoral cutaneous vein

Identify the labeled veins of the lower extremity on the following figure:

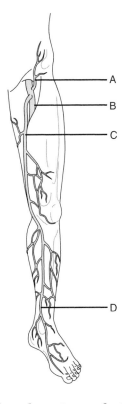

A = Superficial circumflex iliac vein
B = Femoral vein
C = Great saphenous vein
D = Great saphenous vein

What other vein may drain into the great saphenous vein, but is not present in all individuals?

The accessory saphenous vein, a major communicating vein between the great and small saphenous veins

What fascial landmark denotes the junction of the greater saphenous vein with the femoral vein?

The fossa ovalis, an oval-shaped aperture in the fascia lata

What is the function of the perforating (anastomotic) veins of the lower extremity?	They drain the superficial venous system into the deep venous system. Incompetence/dysfunction of these perforating veins contributes to varicose veins of the lower extremities.

INNERVATION

Which cutaneous nerve innervates most of the:	
Anterior and medial thigh?	The femoral nerve
Posterior thigh and popliteal fossa?	The posterior femoral cutaneous nerve (a branch of the sacral plexus)
Cutaneous innervation to much of the lateral thigh is via which nerve?	The lateral femoral cutaneous nerve (a branch of the lumbar plexus)
Which 3 nerves provide cutaneous innervation to the superomedial thigh?	1. Obturator nerve (cutaneous branch) 2. Ilioinguinal nerve 3. Genitofemoral nerve
How does the obturator nerve enter the thigh?	Via the obturator foramen
What is the largest nerve in the body?	The sciatic nerve
How does the sciatic nerve enter the thigh?	Via the greater sciatic notch, below the piriformis muscle
What is sciatica?	Pain in the distribution of a lumbar or sacral nerve root; most often results from nerve compression by a herniated nucleus pulposus (slipped disc) of the lumbar spine
Nerves from which vertebral levels supply fibers to the sciatic nerve?	L4–S3

In general, which nerves supply which compartments of the thigh?	Remember: "**MAP OF S**ciatic" **M**edial Compartment: **O**bturator **A**nterior Compartment: **F**emoral **P**osterior Compartment: **S**ciatic

FEMORAL TRIANGLE

What are the boundaries of the femoral triangle:	Remember: "**So I M**ay **A**lways **L**ove **S**ally"
Superiorly?	The **I**nguinal ligament
Medially?	The medial border of the **A**dductor longus muscle
Laterally?	The medial border of the **S**artorius muscle
What forms the floor of the femoral triangle?	The adductor longus muscle, the pectineus muscle, and the iliopsoas muscle
The roof?	The fascia lata
Name 4 structures that are contained with the femoral triangle.	1. The femoral nerve and its branches 2. The femoral artery and its branches 3. The femoral vein and its tributaries 4. The lymphatics that drain the lower limb (contained within an "empty" space filled with fat and connective tissue) To remember the structures contained within the femoral triangle (from lateral to medial), think **NAVEL**: **N**erve **A**rtery **V**ein **E**mpty space containing **L**ymphatics
What landmark is used to find the femoral vein for placement of a femoral venous line?	The femoral pulse. Aim 1 fingerbreadth *medial* to it.

Describe the location of the femoral artery in the femoral triangle.	The femoral artery is located 2–3 cm inferior to the midpoint of the inguinal ligament (just medial to the femoral nerve). It lies between the vastus medialis and adductor longus muscles.
What is the femoral sheath?	A fascial tube that surrounds the femoral artery, the femoral vein, and the empty space containing the lymphatics within the femoral triangle
The femoral sheath is in continuity with which layer of anterior abdominal wall fascia?	The transversalis fascia
What are the 3 compartments of the femoral sheath and what is contained in each?	1. The lateral compartment (contains the femoral artery) 2. The intermediate compartment (contains the femoral vein) 3. The medial compartment, or femoral canal (the "empty" space containing the lymphatics)
Which structure in the femoral triangle is NOT contained within the femoral canal?	The femoral nerve

ADDUCTOR CANAL

What is the adductor canal?	A fascial tunnel in the thigh located deep to the sartorius muscle
What are the boundaries of the adductor canal: **Laterally?**	The vastus medialis muscle
Posteromedially?	The adductor longus and adductor magnus muscles
Anteriorly?	The sartorius muscle
What forms the roof of the adductor canal?	The sartorius muscle and the subsartorial fascia

What 4 structures pass through the adductor canal?

1. Femoral artery
2. Femoral vein
3. Saphenous nerve
4. Nerve to the vastus medialis

 What is the clinical significance of the adductor canal (aka Hunter's canal?)

It is a site of anatomic narrowing, which leads to turbulent flow. Accordingly, it is a common site for atherosclerotic disease of the lower extremity circulation.

KNEE REGION

KNEE JOINT

Which bones comprise the knee joint?

The femur, tibia, and patella (Note that the fibula does not articulate with the patella.)

Identify the labeled structures on the following anterior view of the dissected knee joint:

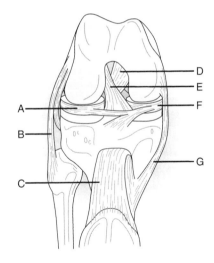

A = Lateral meniscus
B = Fibular (lateral) collateral ligament
C = Patellar ligament
D = Posterior cruciate ligament
E = Anterior cruciate ligament (ACL)
F = Medial meniscus
G = Tibial (medial) collateral ligament (MCL)

Identify the labeled
structures on the following
posterior view of the
dissected knee joint:

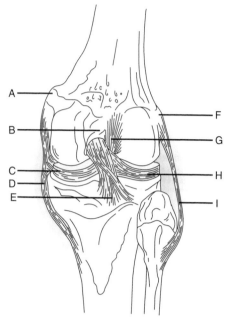

A = Medial femoral epicondyle
B = Intercondylar notch
C = Medial meniscus
D = Tibial (medial) collateral ligament
E = Posterior cruciate ligament
F = Lateral femoral epicondyle
G = Anterior cruciate ligament
H = Lateral meniscus
I = Fibular (lateral) collateral ligament

Which ligament runs from
the lateral femoral
epicondyle to the head of
the fibula?

The fibular (lateral) collateral
ligament. Injury usually occurs with a
"varus" force (from medial to lateral).

Which ligament runs from
the medial femoral
epicondyle to the medial
surface of the tibial shaft?

The tibial (medial) collateral
ligament. Injury usually occurs with a
"valgus" force (from lateral to medial).

Which oblique ligament arises on the:

Posterior tibia and inserts on the medial femoral condyle?

The posterior cruciate ligament

Anterior tibia and inserts on the lateral femoral condyle?

The anterior cruciate ligament

What is the anterior drawer test?

Test for ACL stability/injury; performed by pulling the tibia anteriorly with the knee flexed at 90°. ACL injury is associated with abnormal laxity of the tibia.

Why are the anterior and posterior cruciate ligaments so named?

Because the 2 ligaments cross one another like an "X"

What is the "unhappy triad" of knee injuries?

Injury to ACL, MCL, and medial meniscus; a classic pattern of knee injury caused by a blow to the knee from a lateral direction (i.e., clipping in football)

Describe the relationship of the posterior cruciate ligament to the anterior cruciate ligament

Remember: "**PAM**'s **AP**p**L**es"

Posterior passes **A**nterior inserts
Medially
Anterior passes **P**osterior inserts
Laterally

What is the patellar ligament?

The continuation of the quadriceps tendon from the apex of the patella to the tibial tuberosity

What are the major flexors at the knee joint?

The posterior thigh muscles (i.e., the semimembranosus, semitendinosus, biceps femoris, and hamstring portion of the adductor magnus muscles) and 2 of the posterior leg muscles (i.e., the gastrocnemius and plantaris muscles)

Which muscle is the prime extensor at the knee joint?

The quadriceps femoris muscle

What muscle provides for some rotation at the knee?

The popliteus muscle, a posterior leg muscle that courses from the lateral femoral condyle to the popliteal surface of the tibia. It medially (internally) rotates the knee when near terminal extension to "lock home" the knee joint in extension.

Why can the knee only rotate when it is in the fixed position?

Flexion at the knee allows for relaxation of the collateral ligaments.

Name the 4 bursae of the knee.

1. Suprapatellar bursa
2. Prepatellar bursa
3. Subcutaneous infrapatellar bursa
4. Deep infrapatellar bursa

Which is formed by a superior extension of the synovial membrane of the knee joint?

The suprapatellar bursa (a favorite site for diagnostic joint aspiration)

What is the pes anserinus and what is its clinical significance?

The pes anserinus (means "foot of the goose") is the insertion of the sartorius, gracilis, and semitendinosus muscles on the proximal anteromedial tibia. This is the site of harvesting for anterior cruciate ligament reconstruction surgery when the gracilis and semitendinosus are used as donor tendon grafts.

What is the innervation of the pes anserinus muscles?

Remember the army drill instructor **"SGT FOS"**

Sartorius → **F**emoral Nerve
Gracilis → **O**bturator Nerve
semi**T**endinosus → **S**ciatic

POPLITEAL FOSSA

What is the popliteal fossa?

The diamond-shaped area behind the knee joint

What are the boundaries of the popliteal fossa:	
Posteriorly?	The skin and superficial fascia
Superomedially?	The semimembranosus and semitendinosus muscles
Superolaterally?	The biceps femoris muscle
Inferomedially?	The medial head of the gastrocnemius muscle
Inferolaterally?	The lateral head of the gastrocnemius muscle and the plantaris muscle
Anteriorly?	The posterior surface of the femur, the posterior capsule of the knee joint, and the posterior surface of the tibia superior to the soleus muscle

What does the popliteal fossa contain?

1. Fat
2. The popliteal artery and vein
3. The small saphenous vein
4. Lymphatics and the popliteal lymph nodes
5. The tibial nerve
6. The common peroneal nerve
7. The posterior femoral cutaneous nerve
8. The articular branch of the obturator nerve
9. The popliteus bursa

 What is a Baker's cyst?

A synovial cyst (abnormal accumulation of joint fluid) in the popliteal space; common cause of pain/swelling in the popliteal region

LEG

MUSCLES

Name the 4 muscle compartments of the leg.

1. Superficial posterior
2. Deep posterior
3. Lateral
4. Anterior

The superficial and deep posterior compartments are often grouped together as a single "posterior" compartment.

Posterior Leg Muscles

Name the 7 muscles of the posterior compartment of the leg.

1. Soleus muscle
2. Gastrocnemius muscle
3. Plantaris muscle
4. Popliteus muscle
5. Flexor digitorum longus muscle
6. Flexor hallucis longus muscle
7. Tibialis posterior muscle

Which of these muscles are in the superficial posterior compartment?

The soleus, gastrocnemius, and plantaris

What is the name of the common tendon for the gastrocnemius and soleus muscles?

The tendo calcaneus (Achilles tendon)

Where does the tendo calcaneus insert?

On the posterior surface of the calcaneus, at the tuber calcanei

Which nerve innervates all of the muscles of the posterior compartment of the leg?

The tibial nerve

Soleus Muscle

Origin?

The posterior aspect of the head of the fibula and the soleal line on the tibia

Insertion?

The posterior surface of the calcaneus, via the tendo calcaneus

Action?

Plantarflexes the foot

Gastrocnemius Muscle

How many heads does the gastrocnemius muscle have?

2

What is the origin of the gastrocnemius muscle?	The lateral femoral condyle (lateral head) and the medial femoral condyle (medial head)
Insertion?	The posterior surface of the calcaneus, via the tendo calcaneus
Action?	Plantarflexes the foot, flexes the leg, and raises the heel during ambulation

Plantaris Muscle

Where is the plantaris muscle located?	Deep to the lateral head of the gastrocnemius muscle (i.e., between the gastrocnemius and soleus muscles)
What is the origin of the plantaris muscle?	The lower lateral supracondylar line of the femur
Insertion?	The posterior surface of the calcaneus
Action?	Plantarflexes the foot; flexes the leg

Popliteus Muscle

Origin?	The lateral condyle of the femur and the arcuate popliteal ligament
Insertion?	The posterior surface of the tibia, superior to the soleal line
Action?	Flexes and rotates the leg medially

Flexor Digitorum Longus Muscle

Origin?	The posterior surface of the shaft of the tibia
Insertion?	The bases of the distal phalanges of the lateral 4 toes
Action?	Flexes the lateral 4 toes and plantarflexes the foot

Flexor Hallucis Longus Muscle

Origin?

The posterior surface of the shaft of the fibula

Insertion?

The base of the distal phalanx of the great toe

Action?

Flexes the distal phalanx of the great toe

Tibialis Posterior Muscle

Origin?

The posterior interosseus membrane, the posterior tibia, and the fibula

Insertion?

The navicular tuberosity, cuneiform, and cuboid bones (i.e., all of the tarsal bones except the talus and calcaneus) and the bases of the second, third, and fourth metatarsal bones

Action?

Plantarflexes and inverts the foot

Lateral Leg Muscles

Name the 2 lateral leg muscles.

1. Peroneus longus muscle
2. Peroneus brevis muscle

Which nerve innervates both lateral leg muscles?

The superficial peroneal nerve

Peroneus Longus Muscle

Origin?

The head and upper lateral surface of the fibula

Insertion?

The base of the first metatarsal bone and the medial cuneiform bone

Action?

Everts and plantarflexes the foot

Peroneus Brevis Muscle

Origin?

The lower lateral side of the fibula and the intermuscular septa

Insertion?

Tuberosity at the base of the fifth metatarsal bone

Action?

Everts and plantarflexes the foot

Anterior Leg Muscles

Name the 4 muscles of the anterior compartment of the leg.

1. Tibialis anterior muscle
2. Extensor digitorum longus muscle
3. Extensor hallucis longus muscle
4. Peroneus tertius muscle

Which nerve innervates the anterior leg muscles?

The deep peroneal nerve

Tibialis Anterior Muscle

Origin?

The lateral tibial condyle and the interosseus membrane

Insertion?

The medial cuneiform bone and the first metatarsal bone

Action?

Dorsiflexes and inverts the foot

Extensor Digitorum Longus Muscle

Origin?

The lateral tibial condyle, the proximal two thirds of the fibula, and the interosseous membrane

Insertion?

The bases of the middle and distal phalanges of the lateral 4 toes

Action?

Extends the toes and dorsiflexes the ankle joint

Extensor Hallucis Longus Muscle

Origin?

The medial half of the anterior surface of the fibula and the interosseous membrane

Insertion?

The base of the distal phalanx of the great toe

Action?

Extends the great toe and dorsiflexes and inverts the foot

Peroneus Tertius Muscle

Origin?

The distal third of the fibula and the interosseus membrane

Insertion?

The dorsum of the base of the fifth metatarsal bone

Action?

Dorsiflexes and everts the foot

VASCULATURE

Which artery is the continuation of the femoral artery, and where does it run?

The popliteal artery is a continuation of the femoral artery. It begins at the adductor hiatus and runs inferolaterally through the popliteal fossa.

The popliteal artery bifurcates into which 2 branches at the lower border of the popliteus muscle?

The anterior and posterior tibial arteries (The initial portion of the posterior tibial artery is occasionally referred to as the tibioperoneal trunk.)

Describe the course of the anterior tibial artery.

The anterior tibial artery passes through an opening in the interosseus membrane medial to the fibula, and then becomes the dorsalis pedis artery at the ankle joint.

Describe the course of the posterior tibial artery.

After giving off the peroneal artery, the posterior tibial artery runs distally within the posterior deep compartment, passes behind the medial malleolus (of the tibia), and then terminates by dividing into the medial and lateral plantar arteries in the sole of the foot.

Describe the physical examination of pedal pulses (pulses in the feet).

The posterior tibial pulse can normally be palpated just posterior to the medial malleolus. The dorsalis pedis pulse is found on the midanterior foot.

INNERVATION

The sciatic nerve ends at the popliteal fossa, branching into which 2 nerves?

The tibial nerve and the common peroneal nerve

Describe the course of the tibial nerve in the leg.

It runs between the superficial and deep posterior leg muscles, and then runs along the tibia.

The tibial nerve bifurcates into which 2 nerves at the lower border of the popliteus muscle?

The anterior and posterior tibial nerves

Describe the course of the common peroneal nerve in the leg.

It runs posterior to the fibular head and then wraps around the lateral aspect of the proximal fibula as it branches.

What is the clinical significance of the location of the common peroneal nerve?

The superficial location of the common peroneal nerve as it wraps around the proximal fibula leaves it vulnerable to injury during surgery or from trauma.

What are the branches of the common peroneal nerve?

The superficial peroneal nerve and the deep peroneal nerve

Which muscles are innervated by the superficial peroneal nerve?

The lateral leg muscles (i.e., the peroneus longus and the peroneus brevis muscles)

Which muscles are innervated by the deep peroneal nerve?

The anterior leg muscles (i.e., the tibialis anterior, extensor hallucis longus, extensor digitorum longus, and peroneus tertius muscles) and the extensor digitorum brevis and extensor hallucis brevis muscles of the foot

Injury to the common peroneal nerve leads to what clinical deficit?

Foot drop (inability to dorsiflex the foot)

What nerve innervates the skin on the medial aspect of the leg?

Saphenous nerve

What is a lower-extremity "compartment syndrome"?

A clinical syndrome where swelling within the leg's tight fascial compartments after trauma leads to compromise of circulation to the leg; treated by "fasciotomy," or surgical release of the fascial compartments

What are "shin splints"?

Exercise-induced anterior compartment hypertension; seen in runners

Name each of the 4 compartments of the leg on the figure below, and list the key contents of each:

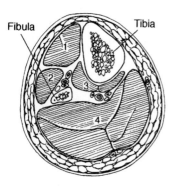

Fibula   Tibia

1. **Anterior compartment**

Muscles: Tibialis anterior, extensor hallucis longus, extensor digitorum longus, and peroneus tertius
Nerve: Deep peroneal
Artery: Anterior tibial

2. **Lateral compartment**

Muscles: Peroneus longus/brevis
Nerve: Superficial peroneal

3. **Deep posterior compartment**

Muscles: Tibialis posterior, flexor hallucis longus, and flexor digitorum longus
Nerve: Tibial
Arteries: Posterior tibial, peroneal

4. **Superficial posterior compartment**

Muscles: Gastrocnemius, soleus, and plantaris

ANKLE REGION

Which bones articulate at the ankle joint?

The tibia, fibula, and talus

Which movements occur at the ankle joint?

Dorsiflexion (toes up) and plantarflexion (toes down) are major movements. Rotation, abduction, and adduction are possible when the foot is plantarflexed.

Which joint provides for most of the foot's ability to invert or evert?

The subtalar joint, which lies between the calcaneus and talus bones of the hindfoot

Which 2 muscles primarily invert the foot (i.e., turn the sole posterior muscles inward)?

The tibialis anterior and the tibialis posterior

Which 3 muscles evert the foot (i.e., turn the sole outward)?

The peroneus longus, peroneus brevis, and peroneus tertius muscles

Which muscles are responsible for dorsiflexion at the ankle?

The extensor hallucis longus and extensor digitorum longus muscles dorsiflex the foot. The tibialis anterior muscle dorsiflexes and inverts the foot.

Which muscles are responsible for plantarflexion at the ankle?

The flexor hallucis longus, flexor digitorum longus, tibialis posterior, gastrocnemius, soleus, and plantaris muscles

Describe the medial and lateral ligaments of the ankle joint.

The medial ligament is deltoid-shaped and is thicker and stronger than the lateral ligament. The lateral ligament consists of 3 bands (the anterior and posterior talofibular ligaments and the middle calcaneofibular band), collectively termed the lateral ligament complex.

What is the most common anatomic injury with an ankle sprain?

The most common mechanism is an "inversion" injury (turning of the foot inward), with resultant stretching or tearing of the "lateral ligament complex" (talofibular ligaments).

Which 2 tendons pass behind the lateral malleolus (of the fibula) at the ankle joint?	The tendons of the peroneus longus and peroneus brevis muscles
Which 2 structures hold these 2 tendons in place?	The superior and inferior peroneal retinacula
What is the tarsal tunnel and what are the structures contained in the tarsal tunnel?	The tunnel formed by the flexor retinaculum posterior and inferior to the medial malleolus. Remember: "**T**om **D**ick **A**nd a **V**ery **N**ervous **H**arry" From anterior to posterior: **T**ibialis anterior tendon Flexor **D**igitorum longus tendon Posterior tibial **A**rtery Posterior tibial **V**ein Tibial **N**erve Flexor **H**allucis longus
What is the clinical significance of the tarsal tunnel?	Compression in the tarsal tunnel can lead to tibial nerve damage resulting in weakness of the plantar foot muscles and numbness of the plantar foot.

FOOT

CUTANEOUS INNERVATION

What nerves provide sensory innervation to the skin of the foot, and what is the distribution of each?	Tibial nerve → sole of the foot Sural nerve → lateral aspect of the foot Saphenous nerve → medial aspect of the foot Deep peroneal nerve → first toe web space Superficial peroneal nerve → dorsum of foot (excluding first toe web space)

DORSAL FOOT MUSCLES

What are the dorsal foot muscles?	The extensor hallucis brevis muscle and the extensor digitorum brevis muscle
What is the action of these muscles?	They extend the toes.

Where do these muscles originate?	On the dorsal surface of the calcaneus
Where does the extensor hallucis brevis muscle insert?	On the base of the proximal phalanx of the great toe
Where does the extensor digitorum brevis muscle insert?	On the base of the proximal phalange of the lateral 4 toes
Which nerve innervates the dorsal muscles of the foot?	The deep peroneal nerve

PLANTAR FOOT MUSCLES

What is the function of the muscles of the sole of the foot?	Toe movement, as well as arch maintenance
The plantar foot muscles are structured into how many layers?	4
What nerves innervate the plantar foot muscles?	The medial and lateral plantar nerves
List the 3 muscles that form the first (superficial) layer of muscles in the plantar foot.	1. Abductor hallucis muscle 2. Abductor digiti minimi muscle 3. Flexor digitorum brevis muscle
List the 2 muscles that form the second layer of muscles in the plantar foot.	1. Quadratus plantae muscle 2. Lumbrical muscles (4)
List the 3 muscles that form the third layer of muscles in the plantar foot.	1. Flexor hallucis brevis muscle 2. Adductor hallucis muscle 3. Flexor digiti minimi brevis muscle
List the 2 groups of muscles that form the fourth layer of muscles in the plantar foot.	1. Plantar interossei (2) 2. Dorsal interossei (4)

What 4 plantar foot muscles are innervated by the medial plantar nerve?

Remember: "**S**ome **L**ive **F**or **H**ot **A**nd **H**appy **D**ancing **B**runettes"
Second toe **L**umbrical
Flexor **H**allucis brevis
Abductor **H**allucis
Flexor **D**igitorum **B**revis
All remaining plantar muscles are innervated by the lateral plantar nerve.

What is plantar fasciitis?

Inflammation of the plantar fascia (aka "heel spurs"); common cause of inferior heel pain; often associated with a history of increase in running/weight-bearing activities

![icon] POWER REVIEW

LOWER EXTREMITY

GLUTEAL REGION

Which 3 gluteal muscles originate from the external surface of the ilium?	The gluteus maximus, gluteus medius, and gluteus minimus muscles
What is the innervation of the:	
Gluteus maximus muscle?	The inferior gluteal nerve
Gluteus minimus and gluteus medius muscles?	The superior gluteal nerve
Which other gluteal muscle is supplied by the superior gluteal nerve?	The tensor fasciae latae
Which 2 gluteal muscles insert into the iliotibial tract?	The gluteus maximus muscle and the tensor fasciae latae muscle
What is the name of the square muscle that runs from the ischial tuberosity to the intertrochanteric crest of the femur?	The quadratus femoris muscle

HIP

Which muscle is the major flexor at the hip joint?	The iliopsoas muscle (a combination of the iliacus and psoas major muscles), 1 of the anterior thigh muscles
Name the external rotators of the hip.	Remember: "**P**lay **G**olf **O**r **G**o **O**n **Q**uaaludes" **P**iriformis **G**emellus superior **O**bturator internus **G**emellus inferior **O**bturator externus **Q**uadratus femoris

Which muscle is the chief extensor of the thigh at the hip joint?	The gluteus maximus muscle

THIGH

Name the contents of the femoral triangle.	**NAVEL** (from lateral to medial): Femoral **N**erve Femoral **A**rtery Femoral **V**ein **E**mpty space containing **L**ymphatics
Which arteries represent the primary arterial supply to the thigh?	The femoral and deep femoral (profunda femoris) arteries
The great saphenous vein empties into which vein?	The femoral vein (at the fossa ovalis)
What are the contents of the adductor canal?	The femoral artery and vein, the saphenous nerve, and the nerve to the vastus medialis
Which muscles comprise the: **Posterior thigh muscles?**	Semimembranosus, semitendinosus, biceps femoris, and adductor magnus (hamstring part) muscles
Medial thigh muscles?	Pectineus, adductor longus, adductor magnus (adductor part), adductor brevis, gracilis, and obturator externus muscles
Anterior thigh muscles?	Iliopsoas, sartorius, and quadriceps femoris muscles
Which nerve innervates the: **Posterior thigh muscles?**	The tibial division of the sciatic nerve
Medial thigh muscles?	The obturator nerve
Anterior thigh muscles?	The femoral nerve

KNEE

What are the major flexors at the knee joint?	1. The semimembranosus, semitendinosus, biceps femoris, and adductor magnus (hamstring portion) muscles (i.e., the posterior thigh muscles) 2. The gastrocnemius and soleus muscles (i.e., the posterior leg muscles)
Which muscle is the prime extensor at the knee joint?	The quadriceps femoris muscle
The popliteal artery bifurcates into which 2 branches at the lower border of the popliteus muscle?	The anterior and posterior tibial arteries

LEG

Name the muscles of the compartments of the leg.	**Superficial posterior leg muscles:** Soleus, plantaris, gastrocnemius **Deep posterior leg muscles:** Popliteus, flexor digitorum longus, flexor hallucis longus, and tibialis posterior **Lateral leg muscles:** Peroneus longus and peroneus brevis **Anterior leg muscles:** Tibialis anterior, extensor digitorum longus, extensor hallucis longus, and peroneus tertius

ANKLE

At the ankle joint, the anterior tibial artery becomes which artery?	The dorsalis pedis artery
What are the structures contained in the tarsal tunnel?	Remember: "**T**om **D**ick **A**nd a **V**ery **N**ervous **H**arry" From anterior to posterior: **T**ibialis Anterior tendon Flexor **D**igitorum Longus tendon Posterior Tibial **A**rtery Posterior Tibial **V**ein Tibial **N**erve Flexor **H**allucis Longus

FOOT

Name the sensory nerves of the foot and where they innervate	Tibial→ sole of the foot Sural → lateral aspect of the foot Saphenous → medial aspect of the foot Deep peroneal → first toe web space Superficial peroneal → dorsum of foot (excluding first toe web space)
The plantar foot muscles consist of how many layers?	4

12

Embryology

GENERAL

What are the 3 basic embryologic periods?	**Zygote stage:** Weeks 1–2 **Embryo stage:** Weeks 3–8 **Fetal stage:** Weeks 9–38
What process during the third week of gestation establishes the germ cell layers?	Gastrulation
What narrow groove forms on the epiblast surface at the beginning of gastrulation?	The primitive streak
Remnants of the primitive streak may persist as what pediatric tumor?	A sacrococcygeal teratoma; has tissue from all 3 germ layers
What structures lie at the primitive streak's cephalad end?	The primitive node and pit
What structures are derived from the ectodermal germ cell layer?	Nervous system, skin, hair/nails, pituitary, adrenal medulla, mammary glands, sweat glands, eye lens/retina, enamel of teeth

What are the 3 germ cell layers?

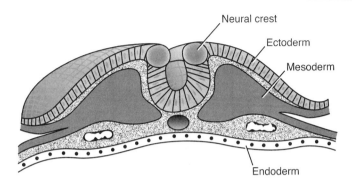

Neural crest
Ectoderm
Mesoderm
Endoderm

1. Ectoderm (outermost)
2. Mesoderm
3. Endoderm (innermost)

What structures are derived from the mesodermal germ cell layer?

Connective tissue, bone, muscle, blood, kidneys, gonads, heart, spleen, adrenal cortex, blood/lymphatic vessels

What are somites?

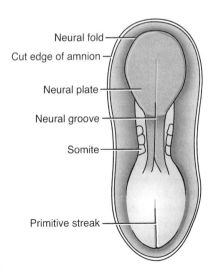

Neural fold
Cut edge of amnion
Neural plate
Neural groove
Somite
Primitive streak

Regionalized paired segments of the "paraxial" mesoderm; lie on either side of the developing neural tube

What is the normal final number of somites?	42–44 pairs
What do somites eventually form?	The *axial* skeleton and associated muscles (Somites do *not* contribute to limb formation!)
What is the origin of the bones and muscles of the limbs?	"Limb buds"; arise from the somatic mesoderm (covered by ectodermal layer)
What structures are derived from the endodermal germ cell layer?	Liver, pancreas, thyroid, parathyroid, and lining of gastrointestinal, genitourinary, and respiratory tracts
What is the first system to begin development after implantation?	The central nervous system (CNS)

BRANCHIAL APPARATUS

What forms the branchial arches (aka pharyngeal arches)?	Bars of thickened mesenchymal tissue adjacent to the foregut, in the region of the future head and neck; contributions from paraxial and lateral plate mesoderm and from neural crests cells
How many branchial arches are there?	While there are initially 6 (numbered from cranial to caudal), the fifth arch regresses without making any significant embryologic contribution.
What are the components of each branchial arch?	An artery, cartilage, a nerve, and muscle tissue
Name the derivatives of cartilage from the:	
First branchial arch.	Malleus/incus, mandible
Second branchial arch.	Stapes, styloid process, hyoid bone (superior body and lesser cornu)
Third branchial arch.	Hyoid bone (inferior body and greater cornu)

Fourth and sixth arches.	Thyroid cartilage, cricoid, arytenoid, and tracheal cartilages

Which nerves develop from the:

First branchial arch?	Maxillary and mandibular divisions of trigeminal nerve (i.e., cranial nerve [CN] V_2 and V_3)
Second branchial arch?	Facial nerve (CN VII)
Third branchial arch?	Glossopharyngeal nerve (CN IX)
Fourth and sixth branchial arches?	Vagus nerve (CN X)

What muscles develop from the:

First branchial arch?	Muscles of mastication, mylohyoid, anterior belly of digastric, tensor tympani, tensor veli palatini
Second branchial arch?	Muscles of facial expression, stapedius, stylohyoid, posterior belly of digastric
Third branchial arch?	Stylopharyngeus
Fourth and sixth branchial arches?	Cricothyroid, levator veli palatini, pharyngeal constrictors, intrinsic laryngeal muscles, esophageal muscle

What internal endoderm-lined recesses develop between the branchial arches?	Branchial (pharyngeal) pouches. The first pouch is between first and second arches.

Name the derivatives of the:

First branchial pouch.	Tympanic membrane, external auditory tube, tympanic cavity
Second branchial pouch.	Palatine tonsils, tonsillar fossa
Third branchial pouch.	Inferior parathyroids and thymus

Fourth/fifth branchial pouch.	Superior parathyroids and parafollicular ("C") cells of thyroid, which form from the "ultimobranchial body" of the fifth pouch
What is the name of the external ectodermal tissue lining each branchial pouch?	Branchial clefts
What are the normal derivatives of the branchial clefts?	The first branchial cleft develops into the external auditory meatus. Branchial clefts 2, 3, and 4 involute.
What is a branchial cleft cyst?	Abnormal branchial cleft remnants; generally treated with excision
Where is a first branchial cleft cyst found?	Just anterior to the ear, close to the parotid gland
Where is a second branchial cleft cyst found?	Along the anterior border of the sternocleidomastoid, anywhere from the angle of the jaw to the clavicle

NERVOUS SYSTEM

The nervous system develops from what part of the embryonic ectoderm layer?	The neural plate
The neural plate differentiates into what 2 structures?	1. Neural tube (becomes central nervous system, i.e., the brain and spinal cord) 2. Neural crest (becomes peripheral nervous system)
The neural tube is formed by the elevation and midline fusion of what structures?	Neural folds (lateral edges of the neural plate)
What is the eventual fate of the neural tube's lumen?	Becomes ventricular system in brain and central canal of spinal cord
What is the fate of the thickened ventral aspect of the neural tube (basal plates)?	Forms the motor aspects of the spinal cord and the brain

The neural tube initially communicates with the amniotic cavity through what 2 openings?

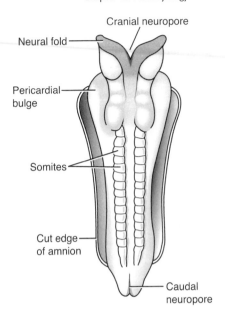

Cranial neuropore

Neural fold

Pericardial bulge

Somites

Cut edge of amnion

Caudal neuropore

The rostral and caudal neuropores, both of which close in fourth week

What about the dorsal thickenings (alar plates)?

Form the sensory aspects of the cord and the brain

What 2 types of processes arise from neuroblasts (primitive nerve cells)?

A primitive axon and primitive dendrites

What is developmentally unique about neuroblasts, relative to other cell types in the body?

Neuroblasts cannot divide or regenerate in adult life. This explains why brain/spinal cord injuries are generally irreversible.

What are neural crest cells?

Ectodermal cells that arise along each edge of the neural folds

Neural crests cells form what cell types?

Neurons and supportive cells of peripheral nervous system; adrenal medulla; "C" cells of the thyroid; pigment cells of the epidermis (melanocytes); skeletal components of the face; meninges; teeth odontoblasts

What cells are responsible for the myelination of peripheral nerves?

Schwann cells (from neural crest)

Myelination of the spinal cord/CNS is performed by what cell type?

Oligodendroglial cells

What is spina bifida?

A category of congenital abnormalities resulting from abnormal/incomplete closure of the neural folds; continuum of defects in the meninges, vertebrae, or overlying skin

What are the 3 dilations of the cephalic neural tube (primary brain vesicles)?

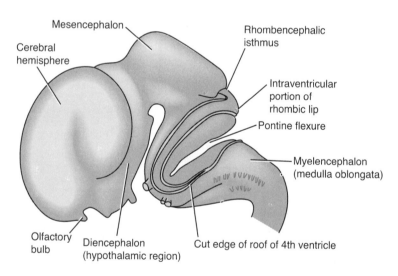

Prosencephalon (forebrain)
Mesencephalon (midbrain)
Rhombencephalon (hindbrain)

The prosencephalon contributes to which parts of the brain?

The cerebral hemispheres, pituitary, thalamus, hypothalamus, cranial nerves (CN) I and II with associated nuclei

What are the 2 major divisions of the prosencephalon?

1. **Diencephalon (posterior):** Forms thalamus, hypothalamus, and pituitary

2. **Telencephalon:** Forms cerebral hemispheres

What are the 2 developmental parts of the hypophysis (pituitary gland)?	1. Rathke's pouch (forms anterior pituitary) 2. Infundibulum (forms posterior pituitary and stalk)
Rathke's pouch may persist into adulthood as what common pituitary tumor?	Craniopharyngioma
The mesencephalon contributes to which parts of the brain?	The midbrain, CN III with its associated nuclei
The rhombencephalon contributes to which parts of the brain?	Pons, cerebellum, medulla oblongata, CN IV–XII with associated nuclei
What is the Arnold-Chiari malformation?	Caudal displacement and herniation of cerebellum through foramen magnum; often associated with other CNS developmental abnormalities
What is the origin of the sympathetic nervous system?	Neural crest cells. Neural crest cells also migrate to form the *medulla* of the adrenal glands.

FETAL CIRCULATION

What is the usual number of umbilical veins?	1 (oxygenated blood)
What is the usual number of umbilical arteries?	2 (deoxygenated blood)
Fifty percent of oxygenated blood bypasses the liver to the inferior vena cava (IVC) through which structure?	Ductus venosus
Oxygenated blood passes from the right atrium to the left atrium through which structure?	Foramen ovale

Blood goes from the pulmonary artery to the descending aorta through which structure?	Ductus arteriosus
What percent of fetal cardiac output normally goes to the lungs?	7%
What factors stimulate the transition from fetal circulation after delivery?	Loss of placental blood flow and beginning of respiration, with associated changes in oxygen tension, pH, and pulmonary vascular resistance

What is the remnant of the:

Umbilical arteries?	Medial umbilical ligaments, internal iliac and superior vesical arteries
Umbilical vein?	Ligamentum teres
Ductus venosus?	Ligamentum venosum
Ductus arteriosus?	Ligamentum arteriosum
Foramen ovale?	Fossa ovalis

CARDIOVASCULAR SYSTEM

What separates the pericardial cavity from the 2 pleural cavities?	The pleuropericardial membranes
What adult structure is a remnant of the pleuroperitoneal membranes?	The fibrous pericardium

HEART

The heart and cardiovascular system are derived from what germ cell layer?	Mesoderm from the "cardiac plate" forms the "heart tube," which undergoes a complex series of bends and folds.
The left and right atria are both derived from what structure?	The primitive atrium

What are the major parts of the heart tube?

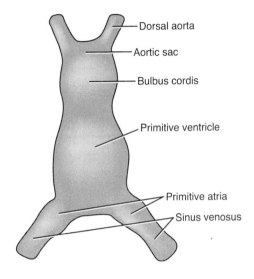

Sinus venosus, primitive atrium, primitive ventricle, and bulbus cordis

The primitive ventricle forms what part of the adult heart?

Left ventricle

What are the 3 parts of the bulbus cordis?

1. The truncus arteriosus
2. Conus cordis
3. Trabeculated portion of the right ventricle

What structures are derived from the truncus arteriosus?

Proximal aorta and pulmonary artery; separated by formation of aorticopulmonary septum

What is the fate of the conus cordis?

It contributes to the outflow tracts of both the right and left ventricles.

What is the fate of the sinus venosus?

It contributes to the wall of the right atrium and the coronary sinus.

The sinoatrial node (pacemaker) is derived from what part of the heart?

The right wall (horn) of the sinus venosus

What part of the heart forms the atrioventricular node and bundle of His?

The left wall of the sinus venosus

What structures form in the fifth week of development to separate the atrioventricular canal into left and right sides?

Endocardial cushions (4)

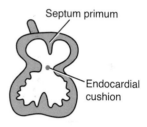

Endocardial cushions go on to form what structures?

The mitral (left) and tricuspid (right) valves, which separate the atria from the ventricles

What structures separate the 2 atria?

Septum primum and septum secundum

What is the name of the initial defect in septum primum?

Ostium primum; later obliterated by fusion of septum primum with endocardial cushions, forming the septum secundum

What is the name of the normal defect in septum secundum?

Foramen ovale; allows blood to pass from the right atrium to the left atrium, bypassing the lungs; normally closes after birth by fusion of septum primum and secundum

What are the 2 parts of the interventricular septum?

Muscular and membranous. Failure of their development/fusion may lead to a ventricular septal defect (VSD), the most common congenital heart abnormality.

VASCULAR SYSTEM

Arteries

How many paired aortic arches are there?	5 (although they are numbered I, II, III, IV, and VI, number 5 regresses!). They form from the branchial arches.
What arteries form from the:	
First arch?	The maxillary artery
Second arch?	The hyoid and stapedial arteries
Third arch?	The common carotid, external carotid, and proximal internal carotid arteries
Fourth arch?	**Left:** Part of aortic arch **Right:** Proximal right subclavian
Fifth arch?	None; it regresses.
Sixth arch?	**Right:** Proximal right pulmonary artery **Left:** Ductus arteriosus
What is an aortic coarctation?	A partial or complete obstruction of the aortic lumen below the origin of the left subclavian artery; may present in adult life as an abnormal difference in blood pressures between left and right extremities
Remnants of the umbilical arteries form what vessels?	Internal iliac and superior vesical arteries
What is the fate of the vitelline arteries?	They form the celiac, superior mesenteric, and inferior mesenteric arteries.

Veins

What are the vitelline veins?	Vessels running between the yolk sac and sinus venosus of the heart

What are the 3 major pairs of veins in the embryo stage?   Vitelline, umbilical, and cardinal. All 3 sets of veins drain into the sinus venosus.

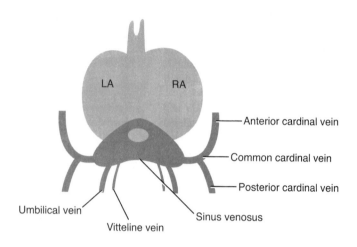

What is the eventual fate of the vitelline vein on the:

Right?   Becomes superior mesenteric vein and proximal portal vein; contributes to IVC

Left?   It disappears.

What is the role of the umbilical veins?   They connect the heart (sinus venosus) with the primitive placenta.

What is the eventual fate of the umbilical vein on the:

Right?   It regresses.

Left?   Forms ligamentum teres, a fibrous cord in the free edge of the falciform ligament

What is the function of the cardinal veins?   Drain the body of the embryo proper

What is the eventual fate of the cardinal vein on the:

Right?   Forms the azygous vein

Left?   It regresses.

LUNGS

The lungs develop from what part of the foregut?	Laryngotracheal groove; forms respiratory lining; joined by mesoderm, which forms associated muscle and cartilage
What outpouching develops off of the foregut at the laryngotracheal groove?	The lung bud; eventually develops into left and right bronchial buds, which divide in stages to form the bronchial tree
What separates the lung bud from the adjacent foregut that will become the esophagus?	Tracheoesophageal folds/septum. Abnormal development may lead to a tracheoesophageal fistula or communication between the trachea and esophagus.
What critical substances are produced by type II pneumocytes?	Surfactant; reduces alveolar surface tension. Inadequate levels in premature infants contribute to the neonatal respiratory distress syndrome.
What percent of alveoli are present at birth?	10%. Development continues until approximately age 8.
What is pulmonary sequestration?	Lung that develops outside the normal investment of pleura, does not communicate with tracheobronchial tree; systemic blood supply with variable drainage

MUSCULOSKELETAL SYSTEM

The musculoskeletal system is derived from which germ cell layers?	Mesoderm (paraxial and lateral plate mesoderm) and ectoderm (neural crest)
The neural crest layer contributes to which part of this system?	The craniofacial skeleton
What are the 2 types of bone formation (ossification)?	1. **Intramembranous** (dermal mesenchyme → bone) 2. **Endochondral** (mesenchyme → cartilage → bone)
Long bones of the extremities form by which type of ossification?	Endochondral

Which bones form by intramembranous ossification?

Flat bones of skull, clavicle

Gaps between the flat bones of the skull are termed what?

Sutures

What is craniosynostosis?

Premature closure of the cranial sutures; leads to deformity of skull

What is the first bone to begin ossification during fetal development but the last to complete it?

The clavicle (Ossification begins at 7 weeks gestation and ends 21 years after birth.)

Describe the rotation of the upper and lower limbs during week 7.

Upper limb rotates laterally ~90°, so extensor muscles end up lateral and posterior.
Lower limb rotates medially ~90°, so extensor muscles end up anterior.

Growth of the limb bones occurs within what region?

The epiphyseal plate of either end (between diaphysis [shaft] and epiphyses [ends])

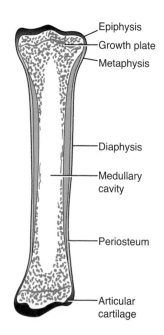

Epiphysis
Growth plate
Metaphysis
Diaphysis
Medullary cavity
Periosteum
Articular cartilage

Which leg bone is analogous to the:	
Radius?	Tibia
Ulna?	Fibula
What is the fate of the primary axial artery of the:	
Arm?	Becomes brachial artery
Forearm?	Becomes common interosseous artery
Thigh?	Becomes profunda femoris artery
Leg?	Becomes anterior and posterior tibial arteries
What is the derivation of the intervertebral discs?	Nucleus pulposus (remnant of notochord) and annulus fibrosis
Incomplete fusion of the bony vertebral arches may lead to what abnormality?	Spina bifida
What are somites?	Segments of the "paraxial" mesoderm that appear from cranial to caudal beginning in the third week of development
Somites form which part of the musculoskeletal system?	Skeletal muscle, cartilage, bone, and subcutaneous tissues
What is the derivation of smooth and cardiac muscle?	Splanchnic mesoderm

ENDOCRINE

ADRENAL GLANDS

What is the embryologic origin of the adrenal glands?	Adrenal cortex from mesoderm Adrenal medulla from neuroectoderm (neural crest cells)

THYROID GLANDS

The thyroid gland descends from what part of the developing foregut?	The floor of the primitive oropharynx, at the base of the tongue

What is the name of the tubular origin of the thyroid gland from the tongue?	The thyroglossal duct; normally obliterates after thyroid descent is complete
What is the name of the embryologic remnant of the thyroid's origin from the tongue?	Foramen cecum
What part of the thyroid is not of endodermal origin?	Thyroid parafollicular ("C") cells; neural crest tissue derived from the fifth pharyngeal pouch
What is a thyroglossal duct cyst?	An abnormal midline remnant of the thyroid's descent in the neck; surgical resection is termed the Sistrunk procedure; includes removal of central hyoid bone
What is a lingual thyroid?	Abnormal thyroid tissue at the base of the tongue, adjacent to foramen cecum; due to failure of normal thyroid descent

PARATHYROID GLANDS

What is the embryologic origin of the superior parathyroid glands?	Fourth pharyngeal pouch
What is the embryologic origin of the inferior parathyroid glands?	Third pharyngeal pouch (with thymus)
What is the normal number of parathyroid glands?	4: 2 superior and 2 inferior; However, 5 glands are present in >5% of people, and ~3% of people have at least 1 absent gland.
What is an "ectopic" parathyroid gland?	Abnormal location of a parathyroid gland; more often occurs with inferior glands, which may even be found in the mediastinum

ABDOMINAL CAVITY

Which structure lies in the free margin of the falciform ligament?

The ligamentum teres

What is the embryologic origin of the ligamentum teres?

The fetal (left) umbilical vein

Which structures form the:
    Median umbilical folds?

A remnant of the urachus

    Medial umbilical folds?

Obliterated fetal umbilical arteries

    Lateral umbilical folds?

Inferior epigastric vessels

DIAPHRAGM

What are the 4 embryologic structures that contribute to the diaphragm?

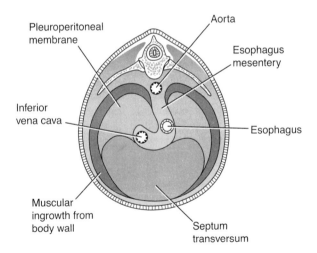

Septum transversum
Pleuroperitoneal folds
Body wall
Esophagus/mesentery

The central tendon of the diaphragm is derived from what structure?

The septum transversum

The pleuroperitoneal folds form what part of the adult diaphragm?

The primitive diaphragm

The mesentery of the esophagus forms which part of the diaphragm?

The crura

What are the 2 major types of congenital diaphragmatic hernias?

1. **Bochdalek:** Posterolateral, L > R, failure of pleuroperitoneal membranes
2. **Morgagni:** Anterior, failure of septum transversum

Primary effect of diaphragmatic hernia is on lung development/function.

ALIMENTARY TRACT

Describe the extent of the:
   Foregut.

From the oropharynx to the hepatopancreatic ampulla (of Vater)

   Midgut.

From the hepatopancreatic ampulla (of Vater) to the distal third of the transverse colon

   Hindgut.

From the distal transverse colon to the anus

What structures are derived from:
   The foregut?

Pharynx, lungs, esophagus, duodenum (up to ampulla), liver, pancreas, bile ducts, gallbladder

   The midgut?

Distal duodenum, jejunum, ileum, cecum, appendix, ascending colon, right two thirds of transverse colon

   The hindgut?

Left one-third of transverse colon, descending colon, sigmoid colon, rectum, superior anal canal

What membrane at the cephalad end of the embryonic foregut normally ruptures in the fourth week?

Buccopharyngeal membrane

Failure of the duodenal lumen to recanalize in week 8 leads to what condition?

Duodenal atresia; presents with neonatal proximal gastrointestinal obstruction, usually at or just distal to ampulla; treatment surgical

Describe the rotation of the stomach that takes place during fetal development.

The stomach rotates 90° **clockwise**; this is why the left vagus nerve is anterior, while the right vagus nerve is posterior.

Describe the rotation of the small bowel.

The small bowel normally rotates 270° **counterclockwise** about the axis of the superior mesenteric artery (SMA).

What is malrotation?

Nonrotation or incomplete rotation of the intestine during development; can occur at a wide range of stages and locations; predisposes midgut volvulus, a life-threatening surgical emergency

What is an omphalocele?

Persistence of the intestines in the umbilical cord (outside abdominal cavity) beyond 12 weeks gestation (normally return into abdominal cavity)

What is a gastroschisis?

Congenital defect in abdominal wall just to the right of the umbilicus, leading to protrusion of intestine and other organs into the amniotic sac

What embryologic structure connects the midgut with the yolk sac?

The vitelline duct (omphalomesenteric duct)

What results from failure of the vitelline duct to obliterate?

Meckel's diverticulum; outpouching from terminal ileum; rule of 2s: present in ~2% of population, ~2 inches long, ~2 feet from ileocecal valve, ~2% of cases are symptomatic (bleeding, inflammation, obstruction)

What is Hirschsprung's disease?	Abnormal migration of ganglionic cells into the rectum, leading to rectal dysfunction; number 1 etiology of pediatric colonic obstruction
The hindgut terminates at what membrane, which normally ruptures in the seventh to ninth week?	Cloacal (anal) membrane. Failure to rupture leads to imperforate anus.
What embryologic structure normally adheres the testicle to the scrotal sac?	The gubernaculum testis. Abnormal development predisposes to testicular torsion, or twisting around its blood supply.

PANCREAS

The pancreas is derived from which part of the gut?	Foregut
The ventral pancreatic bud contributes to which parts of the pancreas?	Inferior head, uncinate process
What are the contributions of the dorsal bud?	Superior head, body, tail
Describe the normal rotation of the pancreatic buds.	As the duodenum elongates, the ventral pancreatic bud migrates inferomedially to join the dorsal bud.
Normal migration and fusion of the pancreatic buds creates which 2 ducts?	1. **Main pancreatic duct (Wirsung):** Most commonly empties into common bile duct 2. **Accessory pancreatic duct (Santorini):** Usually empties into duodenum via minor duodenal papilla
What is pancreatic divisum?	Incomplete fusion of dorsal and ventral pancreatic buds such that the accessory duct (of Santorini) provides the main drainage of the pancreas; present in up to 10% of the population; may lead to chronic abdominal pain or recurrent pancreatitis

What is an annular pancreas?	Head of pancreas circles duodenum as a collar and may constrict lumen; due to failure of ventral bud to rotate properly

LIVER/GALLBLADDER/BILE DUCTS

The liver is derived from what endodermal foregut outgrowth?	The hepatic diverticulum (liver bud)
What are the 2 components of the foregut's hepatic diverticulum?	Pars hepatica and pars cystica
The pars hepatica gives rise to what structures?	Liver, common hepatic duct, intrahepatic bile ducts
The pars cystica gives rise to what structures?	Gallbladder and cystic duct
What is a choledochal cyst?	A congenital dilation of the bile ducts; may be focal or diffuse; etiology unknown
What is biliary atresia?	Congenital closure or absence of the biliary ductal system; patients may present with neonatal jaundice; requires surgical correction; advanced cases treated with liver transplantation

UROGENITAL SYSTEM

The urinary (excretory) and genital (reproductive) systems develop primarily from what cell layer?	Intermediate mesoderm; form from common "urogenital" ridge along the wall of the posterior abdominal cavity

URINARY SYSTEM

Intermediate mesoderm forms which parts of the urinary system?	The kidneys and ureters
What are the 3 embryologic stages of kidney development?	Pronephros → mesonephros → metanephros (adult kidney)

Describe the migration of the kidneys during development.	They begin in the pelvis and ascend into the abdomen until they make contact with the adrenal glands. Migration is due to disproportionate growth between the lumbar and sacral regions.
Kidneys begin to function at what developmental week?	Approximately weeks 11–13; function to recycle swallowed amniotic fluid
What outgrowth of the mesonephric duct forms the collecting system?	The ureteric bud; forms ureter, renal pelvis, calyces, and collecting tubules
What is a horseshoe kidney?	Congenital fusion of the 2 kidneys at their lower pole
The urinary bladder and urethra are derived from what structure?	The urogenital sinus (endoderm); derived from the cloaca, which also forms the anal canal
What is the allantois?	A tubular structure arising from the yolk sac; connects the embryo to the chorion
What is the urachus?	Remnant of the allantois; a fibrous cord that connects the apex of the bladder with the umbilicus; forms the median umbilical ligament

GENITAL SYSTEM

What 2 sets of genital ducts are present in both sexes?	Mesonephric (wolffian) and para-mesonephric (müllerian) ducts
Sex is determined by which chromosome?	The Y chromosome; causes testis to develop
What is the fate of the mesonephric ducts:	
In males?	Become vas deferens, epididymis, and ejaculatory ducts
In females?	They regress.

What is the fate of the paramesonephric ducts:

    In males?   They regress.

    In females?   Become uterus and fallopian tubes

What embryologic defect is associated with the later development of an indirect inguinal hernia?

Patent processus vaginalis; normally closes after testicular descent from the abdominal cavity into the scrotum

What is cryptorchidism?

Abnormal arrest of testicular descent; testicle remains in abdomen or inguinal canal; associated with increased risk for testicular cancer

POWER REVIEW

EMBRYOLOGY

**What are the ADULT
remnants of the following
fetal structures:**

Ductus venosus?	Ligamentum venosum
Umbilical vein?	Ligamentum teres
Umbilical artery?	Medial umbilical ligament, internal iliac and superior vesical arteries
Ductus arteriosus?	Ligamentum arteriosus
Urachus?	Median umbilical ligament
Foramen ovale?	Fossa ovalis
Tongue remnant of thyroid's descent?	Foramen cecum
First branchial pouch?	Tympanic membrane, auditory (eustachian) tube, tympanic cavity
Second branchial pouch?	Palatine tonsils, tonsillar fossa
Third branchial pouch?	Inferior parathyroids and thymus
Fourth/fifth branchial pouch?	Superior parathyroids and parafollicular ("C") cells of thyroid
First branchial cleft?	External auditory meatus
Male mesonephric ducts?	Vas deferens, epididymis, ejaculatory ducts
Female mesonephric ducts?	Regress
Male paramesonephric ducts?	Regress

Female paramesonephric ducts?	Uterus and fallopian tubes
What are the 3 germ cell layers?	1. Ectoderm (outermost) 2. Mesoderm 3. Endoderm (innermost)

What structures are derived from:

Ectoderm?	Nervous system, skin, hair/nails, pituitary, adrenal medulla, mammary glands, sweat glands
Mesoderm?	Connective tissue, bone, muscle, blood, kidneys, gonads, heart, spleen, adrenal cortex, blood/lymphatic vessels
Endoderm?	Liver, pancreas, thyroid, parathyroid, and lining of gastrointestinal, genitourinary, and respiratory tracts
Neural crests cells form what cell types?	Peripheral nervous system; adrenal medulla; "C" cells of thyroid; melanocytes of epidermis; skeletal components of face; meninges; teeth odontoblasts
What are the components of each branchial arch?	An artery, cartilage, a nerve, muscle tissue

What are the 3 dilations of the cephalic neural tube (primary brain vesicles)?	**Prosencephalon (forebrain):** Includes:
	  Telencephalon (cerebral hemispheres)
	  Diencephalon (thalamus, hypothalamus, pituitary)
	Mesencephalon (midbrain)
	Rhombencephalon (hindbrain):
	  Forms pons, cerebellum, medulla oblongata

What are the major parts of the heart tube and the derivatives of each?	**Sinus venosus:** Wall of right atrium, sinoatrial node, atrioventricular node, bundle of His **Primitive atrium:** Both atria **Primitive ventricle:** Left ventricle **Bulbus cordis:** Proximal aorta/pulmonary artery, trabeculated right ventricle, both ventricular outflow tracts

What structures separate the atrioventricular canal into left and right sides?	Endocardial cushions; eventually form the mitral and tricuspid (right) valves
What structures separate the 2 atria?	Septum primum and septum secundum
What is the name of the normal defect in septum secundum?	Foramen ovale; allows blood to pass from the right atrium to the left atrium; normally closes after birth
What are the 3 major pairs of veins in the embryo stage?	Vitelline, umbilical, and cardinal. All 3 sets of veins drain into the sinus venosus.
The lungs develop from what part of the foregut?	Laryngotracheal groove; forms respiratory lining; joined by mesoderm, which forms associated muscle and cartilage
What are the 2 types of bone formation (ossification)?	1. **Intramembranous** (dermal mesenchyme → bone) (e.g., flat skull bones) 2. **Endochondral** (mesenchyme → cartilage → bone) (e.g., long bones of extremity)
What is the first bone to begin ossification during fetal development but the last to complete it?	The clavicle (Ossification begins at 7 weeks gestation and ends 21 years after birth.)
What is the embryologic origin of the adrenal glands?	Cortex from mesoderm Medulla from neuroectoderm (neural crest)
What part of the thyroid is not of endodermal origin?	Thyroid parafollicular ("C") cells; neural crest tissue derived from fifth branchial pouch
What is the embryologic origin of the:	
Superior parathyroid glands?	Fourth pharyngeal pouch
Inferior parathyroid glands?	Third pharyngeal pouch (with thymus)

What are the 4 embryologic structures that contribute to the diaphragm?

1. Septum transversum
2. Pleuroperitoneal folds
3. Body wall
4. Esophagus/mesentery

Describe the extent of the:

Foregut.

Oropharynx to hepatopancreatic ampulla

Midgut.

Hepatopancreatic ampulla to distal third of transverse colon

Hindgut.

Distal transverse colon to anus

What structures are derived from:

The foregut?

Pharynx, lungs, esophagus, duodenum (up to ampulla), liver, pancreas, bile ducts, gallbladder

The midgut?

Distal duodenum, jejunum, ileum, cecum, appendix, ascending colon, right two thirds of transverse colon

The hindgut?

Left one third of transverse colon, descending colon, sigmoid colon, rectum, superior anal canal

Describe the developmental rotation of the:

Stomach.

Rotates 90° **clockwise**

Small bowel.

Rotates 270° **counterclockwise** about the superior mesenteric artery (SMA).

What results from failure of the vitelline duct to obliterate?

Meckel's diverticulum; outpouching from terminal ileum

The ventral pancreatic bud contributes to which parts of the pancreas?

Inferior head, uncinate process

What are the contributions of the dorsal bud?

Superior head, body, tail

The liver is derived from what endodermal foregut outgrowth?

The hepatic diverticulum (liver bud); composed of pars hepatica and pars cystica

The pars hepatica gives rise to what structures?

Liver, common hepatic duct, intrahepatic bile ducts

The pars cystica gives rise to what structures?

Gallbladder and cystic duct

What defect is associated with an indirect inguinal hernia?

Patent processus vaginalis

Appendix — Surgical Anatomy Pearls

GENERAL SURGERY

APPENDECTOMY

Name the tissue layers encountered during an appendectomy incision, from superficial to deep.	Skin, fatty tissue ("Camper's fascia"), Scarpa's fascia, external oblique aponeurosis, internal oblique muscle, transversus abdominis muscle, transversalis fascia, peritoneum
What anatomic landmark universally marks the location of the appendix?	The confluence of the colon's 3 taeniae coli
Where is McBurney's point?	Two-thirds the distance along a line between the umbilicus and anterior superior iliac spine (ASIS); classic location for pain with appendicitis
What is the blood supply to the appendix?	Aorta → superior mesenteric artery → right colic artery → ileocolic artery → appendiceal artery

LAPAROSCOPIC CHOLECYSTECTOMY

What are the borders and contents of Calot's triangle?	Inferior liver edge, common hepatic duct, and cystic duct. This triangle usually contains the cystic artery and a prominent lymph node ("Calot's node").
How may bile enter the gallbladder, besides the cystic duct?	Ducts of Luschka; drain bile from liver directly into gallbladder; may be responsible for bile leak after cholecystectomy
What are the 3 structures that run in the hepatoduodenal ligament and what is their most common relative orientation?	1. Portal vein (posterior) 2. Hepatic artery (left anterior) 3. Common bile duct (right anterior)

HERNIA

An indirect inguinal hernia results from what embryologic event?	Failure of processus vaginalis to close
The inguinal ligament is formed from what abdominal muscle layer?	Inferior edge of external oblique; runs from ASIS to pubic tubercle
What are the borders of and what is the significance of Hesselbach's triangle?	Inguinal ligament, inferior epigastric vessels, and lateral border of rectus muscle. Direct inguinal hernias represent a defect in the inguinal floor (transversalis fascia) of this region.
What landmark differentiates a direct from an indirect inguinal hernia?	Indirect hernias occur through the internal inguinal ring, lateral to the inferior epigastric vessels; direct hernias are medial to the vessels.
What are the borders of the femoral canal (site of femoral hernia)?	**Medial:** Lacunar ligament **Superior:** Inguinal ligament **Lateral:** Femoral vein **Inferior:** Cooper's ligament
What nerve generally runs just anterior to the spermatic cord, and what deficit results from its injury?	Ilioinguinal nerve (numbness of scrotum/medial thigh)
What are the contents of the spermatic cord? How about in females?	Cord is invested by cremasteric muscle fibers (continuation of internal oblique muscle layer) and contains vas deferens/associated vessels, testicular vessels, genital branch of genitofemoral nerve, lymphatics, and autonomic nerves. Equivalent structure in female is the round ligament (usually divided during hernia repair).

COLECTOMY

What is removed in a "right hemicolectomy"?	Terminal ileum (5–10 cm), cecum, ascending colon, hepatic flexure, and variable amount of transverse colon

What anatomic landmark defines the end of the sigmoid colon and the beginning of the rectum?	The splaying out of the 3 taeniae coli
Describe the rectum's blood supply.	Aorta → inferior mesenteric artery → superior rectal artery Common iliac artery → internal iliac artery → middle rectal artery Common iliac artery → internal iliac artery → internal pudendal artery → inferior rectal artery
What is the marginal artery of Drummond? What about the arc of Riolan?	The marginal artery of Drummond is a continuous arterial chain linking the vascular arcades of the entire colon. The arc of Riolan is a collateral channel in the proximal colonic mesentery linking the superior mesenteric artery (SMA) and inferior mesenteric artery (IMA) blood supplies.
What is the location/ significance of Griffith's point? Sudeck's point?	Griffith's point is located at the splenic flexure, at the junction of the SMA and IMA vascular territories. Sudeck's point is at the junction of the sigmoid colon and rectum. These represent "watershed" regions, which are theoretically vulnerable to ischemia (lack of blood supply).

HEMORRHOIDECTOMY

What landmark differentiates internal from external hemorrhoids?	The dentate line. Because internal hemorrhoids occur above this line, they are painless.
What are the locations of the 3 columns of internal hemorrhoidal tissue?	Left lateral, right anterior, and right posterior
Describe the innervation of the anal sphincter muscles.	**Internal sphincter:** Under involuntary control via pelvic autonomic nerves **External sphincter** (composed of three parts): Under voluntary control; more important for bowel control than internal

SPLENECTOMY

What organ lies in close proximity to the spleen and may be inadvertently injured with clamping of the hilar vessels?	Tail of pancreas
What is the blood supply to the spleen?	Splenic artery (largest branch of celiac trunk)

MASTECTOMY

What anatomic landmarks define the extent of breast tissue removed during mastectomy?	**Superior:** Clavicle **Medial:** Sternum **Inferior:** Inframammary crease **Lateral:** Latissimus dorsi
What are the borders of the axilla?	**Medial:** Chest wall and serratus anterior muscle **Lateral:** Humerus **Superior:** Axillary vein **Posterior:** Subscapularis and latissimus dorsi muscles **Anterior:** Pectoralis major and minor muscles
What 2 motor nerves are particularly vulnerable to injury during axillary lymph node dissection and what are the associated deficits?	1. **Long thoracic nerve** (to serratus anterior): Scapular winging 2. **Thoracodorsal nerve** (to latissimus dorsi): weakness of arm extension/adduction
What named lymph nodes occur between the pectoralis major and minor muscles?	Rotter's nodes

THYROID/PARATHYROID

Describe the thyroid's blood supply.	External carotid artery → superior thyroid artery Subclavian artery → thyrocervical trunk → inferior thyroid artery Thyroid ima in <10% of populaion; from aortic arch or brachiocephalic artery

All 3 thyroid veins (superior, middle, inferior) generally drain into internal jugular vein.

What important nerve runs with the inferior thyroid artery and what deficit results from its injury?

The recurrent laryngeal nerve; unilateral injury → hoarseness; bilateral injury → airway obstruction

What nerve runs with the superior thyroid artery?

Superior laryngeal nerve (external branch). Injury may lead to weakening of voice; the injury is not noticed by most, unless patient is a singer.

What is the ligament of Berry?

Posterior suspensory ligament that attaches the thyroid to the cricoid cartilage

What blood vessel constitutes the major blood supply to the parathyroid glands?

The inferior thyroid artery

Describe the location of the superior and inferior parathyroid glands, relative to the inferior thyroid artery.

Superior glands: Superior and posterior to vessel
Inferior glands: Inferior and anterior to vessel

Describe the embryology of the parathyroids.

Superior glands from fourth pharyngeal pouch (with thyroid parafollicular cells)
Inferior glands from third pharyngeal pouch (with thymus)

What percentage of people have only 3 parathyroids? What percentage have 5?

10% have 3; 5% have 5

OTOLARYNGOLOGY

TONSILLECTOMY

What 3 sets of paired lymphoid tissues form "Waldeyer's ring"?

The lingual tonsils (anterior), palatine tonsils (lateral), and pharyngeal tonsils (posterosuperior)

What 3 muscles contribute to the tonsillar fossa?

Palatoglossus muscle, palatopharyngeal muscle, and superior constrictor of pharynx

What nerve lies in close proximity to the tonsillar bed and is therefore vulnerable to injury? What is the associated deficit?	The glossopharyngeal nerve (cranial nerve [CN] IX); associated with loss of sensation on the posterior surface of the tongue
What 3 major arteries lie in close proximity to the tonsillar fossa?	The internal carotid artery, lingual artery, and external maxillary artery
The tonsil receives its primary blood supply from what branches of the external carotid artery?	Ascending pharyngeal, lingual, and facial

CERVICAL LYMPH NODE DISSECTION

What are the borders of the "anterior triangle"?	Anterior margin of sternocleidomastoid, inferior margin of mandible, and vertical midline
What are the borders of the "posterior triangle"?	Borders of the sternocleidomastoid and trapezius muscles and middle third of clavicle
What important neurovascular structures lie within the posterior triangle?	Brachial plexus, subclavian vessels, spinal accessory nerve (CN XI), transverse cervical nerve
What neck structures are preserved in a "modified radical" as opposed to a "radical" neck dissection?	Internal jugular vein, sternocleidomastoid muscle, CN XI
What is Erb's point?	Midpoint along posterior sternocleidomastoid (1 cm superior to greater auricular nerve); used to locate spinal accessory nerve (CN XI)
Where does CN XI exit the skull?	Jugular foramen

What is the origin of the ansa cervicalis? What are the consequences of its surgical interruption?	The ansa appears to arise from the hypoglossal nerve (CN XII), although CN XII does not contribute fibers to it. The ansa can usually be sacrificed without any clinical sequelae.

TRACHEOSTOMY

What are the anatomic landmarks for an emergent surgical airway (cricothyroidotomy)?	Cricoid cartilage (below) and thyroid cartilage (Adam's apple, above)
Delayed bleeding after tracheostomy is what until proven otherwise?	Fistula from innominate artery into trachea
What nerves innervate the larynx?	Superior and recurrent laryngeal branches of vagus nerve (CN X)

PAROTIDECTOMY

Describe the parotid's route of glandular drainage?	Parotid duct ("Stensen's") enters oropharynx adjacent to the second upper molar
What structure separates the parotid gland into superficial and deep lobes?	The facial nerve
What techniques are used to locate the facial nerve during parotid surgery?	The nerve may be located at the stylomastoid foramen, after which the nerve is traced distally. Alternatively, a distal branch can be traced proximally to the origin.
What are the motor branches of the facial nerve (CN VII)?	Temporal, zygomatic, buccal, mandibular, and cervical
Which facial nerve branch is most vulnerable to injury during parotid surgery?	The mandibular branch

Injury to what structures may lead to Frey's syndrome (gustatory sweating)?	Autonomic nerve fibers. Frey's syndrome occurs when misdirected regenerating parasympathetic nerve fibers invade the severed sympathetic fibers that supply sweat glands of the skin.

UROLOGY

PROSTATECTOMY

From what embryologic structure does the prostate arise?	Urogenital sinus
What vessel provides the majority of the prostate's blood supply?	The inferior vesical artery
Most prostate cancers arise in what zone of the gland?	Peripheral
Benign prostatic hypertrophy (BPH) arises in what zone of the gland?	Transitional zone (surrounds the urethra)
What is the most common site of lymphatic spread of prostate cancer? Most common hematogenous site?	**Lymphatic:** Obturator lymph nodes **Hematogenous:** bone
Where do the nerves for erection lie in relation to the prostate?	The neurovascular bundle lies posterolateral to the prostate.

NEPHRECTOMY

Which structures are contained within Gerota's fascia?	Kidney, perirenal fat, adrenal gland, ureter, gonadal vessels
What is the relationship of the renal hilar structures?	From anterior to posterior: Renal vein, renal artery, renal pelvis

ADRENALECTOMY

Describe the blood supply to the adrenal gland.	Superior adrenal artery from inferior phrenic artery Middle adrenal artery from aorta Inferior adrenal artery from renal artery
Why is adrenalectomy more technically challenging on the right than the left?	The right adrenal vein is much shorter and empties directly into the inferior vena cava (the left adrenal vein empties into the left renal vein).

ORCHIECTOMY

What is the arterial supply to the testis?	Internal spermatic artery, cremasteric artery, and artery of the vas
Where is the epididymis in relation to the testicle?	On the testicle's posterolateral surface
What anatomic abnormality predisposes testicular torsion?	"Bell clapper deformity": Transverse lie of the testicle in the scrotum (testicle on its side)
Why is the surgical incision for a radical orchiectomy (for cancer) performed in the inguinal region and not in the scrotum?	So as not to alter the lymphatic drainage of the testicle to the retroperitoneal lymph nodes. Using a scrotal incision could facilitate tumor drainage into the superficial inguinal nodes.

OBSTETRICS/GYNECOLOGY

HYSTERECTOMY

What genitourinary structure is particularly vulnerable to injury during hysterectomy?	The ureter
What are the 3 most common sites of ureter injury during hysterectomy?	1. The pelvic brim when clamping the infundibulopelvic (suspensory) ligament 2. The parametrium when clamping the uterine artery 3. The apex of the vagina when closing the vaginal cuff

What is the origin of the uterine artery?	Internal iliac artery
The bifurcation of the common iliac artery overlies what muscle?	The psoas
The uterine artery crosses the parametrium through which structure?	The cardinal ligament
What is the relationship of the ureter to the uterine artery as they course through the parametrium?	The artery is anterior-superior to the ureter ("bridge over water"). These 2 structures lie in close proximity just superior to the ischial spine.
What structure contains the ovarian vessels?	The infundibulopelvic ligament (superolateral edge of broad ligament)
What structure holds the uterus in anteversion?	The round ligament

CAESEREAN SECTION

What are the 3 structures at the cornua ("horns") of the uterus from anterior to posterior?	The fallopian tube, the round ligament, and the uteroovarian ligament (aka "proper ovarian ligament," runs from uterus to ovary)
What is the consequence of bilateral internal iliac ligation for massive uterine bleeding?	Generally none. There is extensive collateral blood supply to the pelvis via the ovarian and inferior mesenteric arteries.

CARDIOVASCULAR SURGERY

CAROTID ENDARTERECTOMY (CEA)

What are the 8 branches of the external carotid artery, from proximal to distal?	Superior thyroid, ascending pharyngeal, lingual, facial, occipital, posterior auricular, maxillary, and superficial temporal
What are the branches of the internal carotid artery in the neck?	There are none! First branch is ophthalmic (in cranium).

What vein is usually divided to allow better exposure of the carotid bifurcation?

The common facial vein

What is the usual orientation of structures within the carotid sheath?

Carotid artery medial, internal jugular vein lateral, vagus nerve posterior

What cranial nerves are most vulnerable to injury during CEA and what are the associated deficits?

Vagus nerve (CN X): Hoarseness
Hypoglossal nerve (CN XII): Ipsilateral deviation of tongue
Glossopharyngeal nerve (CN IX): Hoarseness/dysphagia
Mandibular branch of the facial nerve (CN VII): inability to raise corner of mouth

ABOMINAL AORTIC ANEURSM (AAA)

What external landmark predicts the usual location of the aortic bifurcation?

The aortic bifurcation lies just above and to the left of the umbilicus.

What branches of the aorta define the level of an AAA?

The renal arteries

What blood vessel generally runs just anterior to the neck of an aortic aneurysm?

The left renal vein (important landmark)

The origin of what anterior branch of the aorta lies in the middle of most AAAs?

The inferior mesenteric artery (IMA); usually sacrificed with AAA repair

What is the most common site of aortoenteric fistula?

Third portion of duodenum; may lead to massive upper gastrointestinal bleed

CORONARY ARTERY BYASS GRAFT (CABG)

What are the 3 coronary arteries?

Left anterior descending ("LAD"; septal and diagonal branches), left circumflex ("circ"; obtuse marginal branches), and right coronary artery (acute marginal branches and posterior descending branch)

| What coronary artery supplies the sinoatrial node? The atrioventricular node? | Both are usually supplied by the right coronary artery. |
| What blood vessel is usually grafted to the LAD? | The left internal mammary artery (LIMA) |

THORACIC SURGERY

PULMONARY LOBECTOMY

How many lobes and segments are there in each lung?	**Left:** Upper lobe (3 segments), middle lobe (2 segments), lower lobe (5 segments) **Right:** Upper lobe (5 segments, including the "lingula"), lower lobe (5 segments)
Which nerve runs anterior to the hilum? Posterior to the hilum?	**Anterior:** Phrenic nerve **Posterior:** Vagus nerve
What is Horner's syndrome and what is its etiology?	Ptosis (eyelid droop), miosis (constricted pupil), anhidrosis (lack of perspiration); from upper lobe lung cancer invading cervical sympathetic chain

Figure Credits

Figure 2-2.
Redrawn from Agur AM, Dalley AF II. Grant's Atlas of Anatomy. 11th Ed. Baltimore: Lippincott Williams & Wilkins, 2005. Fig. 7.1A.

Figure 2-5.
Redrawn from Agur AM, Dalley AF II. Grant's Atlas of Anatomy. 11th Ed. Baltimore: Lippincott Williams & Wilkins, 2005. Fig. 7.4B.

Figure 2-7.
Redrawn from Agur AMR. Grant's Atlas of Anatomy. 9th Ed. Baltimore: Williams & Wilkins, 2001. Fig. 7.43.

Figure 2-10.
Redrawn from Moore KL. Clinically Oriented Anatomy. 3rd Ed. Baltimore: Williams & Wilkins, 1992. Figure 7-83.

Figure 2-16.
Redrawn from Moore KL, Dalley AF II. Clinically Oriented Anatomy. 4th Ed. Baltimore: Lippincott Williams & Wilkins, 1999. Table 7.8B.

Figure 4-3.
Redrawn from Moore KL. Clinically Oriented Anatomy. 3rd Ed. Baltimore: Williams & Wilkins, 1992. Figure 9-3C.

Figure 5-1.
Redrawn with permission from Netter FH. Atlas of Human Anatomy. 3rd Ed. Teterboro, NJ: Icon Learning Systems, LLC, 2003. A subsidiary of MediMedia, USA, Inc. All rights reserved.

Figure 5-3.
Redrawn from Moore KL. Clinically Oriented Anatomy. 3rd Ed. Baltimore: Williams & Wilkins, 1992. Fig. 8-21.

Figure 7-4.
Redrawn from Agur AMR. Grant's Atlas of Anatomy. 9th Ed. Baltimore: Williams & Wilkins, 2001. Fig. 6.40, p. 386.

Figure 7-12.
Redrawn with permission from Marieb EN, Mallatt J. Human Anatomy. 3rd Ed. San Francisco: Benjamin Cummings, 2001. Fig. 26.18b.

Figure 9-4.
Redrawn with permission from Perry CW, Phillips BJ. Rectus sheath hematoma: review of an uncommon surgical complication. Hosp Physician 2001;35–37. Fig. 1. Copyright 2001 by Turner White Communications, Inc.

Figure 10-11.
Redrawn with permission from Netter FH. Atlas of Human Anatomy. 2nd Ed. East Hanover, NJ: Novartis Medical Education, 1997. Copyright 1997, Icon Learning Systems, LLC. A subsidiary of MediMedia, USA, Inc. All rights reserved.

Figure 10-12.
Redrawn from Clemente CD, ed. Gray's Anatomy. 30th American edition. Philadelphia: Lea & Febiger, 1985. Fig. 12-55.

Figure 12-1.
Redrawn from Sadler TW. Langman's Medical Embryology. 9th Ed. Baltimore: Lippincott Williams & Wilkins, 2004. Fig. 5.3A.

Figure 12-2.
Redrawn from Sadler TW. Langman's Medical Embryology. 9th Ed. Baltimore: Lippincott Williams & Wilkins, 2004. Fig. 19.1B.

Figure 12-3.
Redrawn from Sadler TW. Langman's Medical Embryology. 9th Ed. Baltimore: Lippincott Williams & Wilkins, 2004. Fig. 19.3B.

Figure 12-4.
Redrawn from Sadler TW. Langman's Medical Embryology. 9th Ed. Baltimore: Lippincott Williams & Wilkins, 2004. Fig. 19.17.

Figure 12-7.
Redrawn with permission from the Heart Diseases in Children Information Center, University of Chicago, http://pediatriccardiology.uchicago.edu.

Figure 12-9.
Redrawn from Sadler TW. Langman's Medical Embryology. 9th Ed. Baltimore: Lippincott Williams & Wilkins, 2004. Fig. 10.6C.

INDEX

In this index, *see also* cross-references designate related topics or more detailed lists of subtopics